MADUKKA

About the Author

Julie Janson is a Burruberongal woman of the Darug Aboriginal nation. A novelist, playwright, and poet. She began writing plays while living in remote Northern Territory Aboriginal communities making giant puppets, masks and costumes. Her career as a playwright includes: 'Black Mary' Belvoir St Theatre and the Festival of the Dreaming 1997. She was the recipient of the Oodgeroo Noonuccal Poetry Prize 2016 and Judith Wright Poetry Prize 2019. Julie's historical novel *Benevolence* was published by Magabala in 2020 and HarperVia USA, UK in 2022. *Madukka the River Serpent* is her first Indigenous crime novel.

MADUKKA

The River Serpent

UWA PUBLISHING

First published in 2022, reprinted 2022 and 2023 by
McPherson's Printing Group

Crawley, Western Australia 6009
www.uwap.uwa.edu.au

UWAP is an imprint of UWA Publishing
a division of The University of Western Australia

THE UNIVERSITY OF
WESTERN
AUSTRALIA

ISBN: 978-1-76080-229-5

NATIONAL LIBRARY OF AUSTRALIA

A catalogue record for this book is available from the National Library of Australia

Cover design by Lisa White
Typeset in 12 point Dante by Lasertype
Printed by McPherson's Printing Group

uwapublishing

MIX
Paper | Supporting responsible forestry
FSC® C001695
FSC www.fsc.org

I dedicate this novel to the wonderful Aboriginal Aunties who inspired me and shared their wisdom and cultural knowledge: Evelyn Bates, Veronica Saunders, Emily Walker, Jenny Ebsworth, Rachel Carney, Lillian Crombie, Thelma Thorne (dec.) Delma Davison (dec.) and Robyn Williams (dec.)

CHAPTER 1

26th January, Survival Day

A crow picking at eye sockets in a sheep skull was a reminder of how outback country towns were shrivelling up. A desert landscape shone red and grey against the Darling River. The dusty desolation of Ngiyampaa country. Benevolent colonialism. A town beside a stinking green river of dead fish. The ash-filled streets smelt of bushfire, the air heady with barbequed animals. A town where most of the shops had iron-mesh grilles, chains and big locks.

Aunty June had completed her TAFE course in Investigative Services; she proudly showed the Certificate III in a nice frame on the wall. It had taken thirty hours, online. Her office was in the main street; it used to be Vince's Video Rentals and people still dropped overdue items down the hatch. The latest had been *Midnight Cowboy*.

She cleaned the window with a Wettex and swept the footpath out front. The handwritten sign stuck on the window announced: Yanakirri Investigation Services – Confidentiality Guaranteed. The name meant 'to run fast', as most people she chased were doing that. There were no clients except occasional Murri women wanting June to track cheating husbands or men who had run off to escape child support payments. The women didn't pay much – nothing in fact, except the occasional kangaroo leg still covered in grey fur. To illustrate her thoughts, when June opened the fridge just now to get some milk, blood dripped onto the lino floor.

She locked the office door and stepped outside to take in some hot air.

Many other businesses stood empty along the road, most with 'For Sale' signs. The failing economy had gutted the biggest hotel; it sat there now with blackened walls, graffiti and glassless windows. Next door to June's office sat the original Commonwealth Bank, a majestic sandstone building [?] displaying a fly-blown notice on the door: 'This bank has been closed to improve customer service.'

The local gunjie, Police Sergeant Bruce Blackett, pulled up outside June's office in his wagon and sneered. He was in his sixties, with a paunch and thinning grey hair over skin cancers. His black R. M. Williams riding boots were non-uniform issue: convenient for kicking someone. Blackett admired himself in the rear-view mirror and picked his teeth with a match.

June stuck out her chin and oozed annoyance while the copper looked over her broad golden face, pink lips, tall frame and skinny legs. She knew she was a healthy-looking fifty-year-old beauty with white teeth. A freshwater Gamilaraay Aboriginal woman, born of the clay plains, dust and kangaroos.

Blackett pointed at the office sign and hooted; he hung his arm out the window and slapped the driver's door. He sounded very amused. 'Investigated any major crimes lately? Like a murder or diamond heist? Or just some Murri kid pinching a packet of

Arnott's Custard Creams?' said Blackett. 'Cert three in investigative services. Jesus wept.'

June held up a middle finger. He drove off laughing.

* * *

Later in the day, June watched cousin Thommo drive past her white fibro house in a backstreet of Wilga. She dug a hole with a sharp spade in the front garden of limp geraniums. She was aiming to go to the Survival Day concert on the Reserve. The town council was putting on a sausage sizzle in the main street to mark Australia Day and 1788 – the invasion by English colonisers. Some townies dressed up in colonial dresses and danced the polka in the main road. Sticking it to the Blacks all over again.

Thommo the Murri activist turned the corner in his beaten-up Toyota. June beckoned, and he looked unsure, but he parked on the verge and walked over. His dreadlocks hung around his ears and his marijuana-leaf earrings swung in the sun. She loved him like a son. He looked tired and edgy as they leant together.

'I'm coming with you, cuz,' she whispered.

'Too dangerous, Aunt.' He stared at her feet. 'You stay home. Let us young fellas take risks. I got a call from that Steve Allunga, working for the Aboriginal Legal Service in the city. He wrote some articles for *IndigenousX* about stuff like water consortiums corruption up here. The federal government got no water at all from the $80 million it spent. No tender was put out, they just bought it from a fella who is high up in government. Steve wants to talk to you when he gets here.'

'I just take care of divorces and stuff.'

June suspected Thommo's enviro group was planning something illegal. The police arrested him up north a month ago for chaining himself to a trainload of coal. She lay awake at night terrified that he had mad plans, the kind that got you sent to prison for twenty years as a terrorist.

'You're not planning something really stupid, are you?' June grabbed his arm, and he flinched, full of nervous energy. She wanted to take his dear head and hug it to her chest to stop him doing something crazy. June had cared for this little cousin down by the river. Taught him to swim and read. It was the bond of love that took care of her childlessness and history of family pain.

Thommo had an eager smile. He joined every local environmental protest. Fearless fella. Once, he had stayed aloft in a threatened tree, eating baked beans and biscuits, sleeping along a tree limb. He was a bit wild. Women liked him and treated him like a pet.

She watched him pace the garden. His long legs and broad shoulders were evidence of her uncle's handsome build – the uncle who'd disappeared one day when fighting a big mountain fire.

Thommo pulled his jacket around his shoulders. It was a Vinnie's second-hand leather jacket with studs rusting on the shoulders. It made him seem tougher than his usual Aboriginal-flag t-shirt with 'Stop Adani' emblazoned across the back.

A terrible uneasiness filled June's head.

'I respect you, Aunt,' Thommo said, 'but you're not getting into trouble because of me. I had a bit of an encounter with a bad fella. I wanted to get explosives. For an action and then … an ice dealer tried to force me into selling his drugs. Argued at the pub, but Arthur threw me out. I was real nervous.'

'Don't do that rubbish.'

'Yeah, I'm shit-scared. A lot of Murri women look up to you. Keep out of it. I'm just meeting a contact. He marked the spot on my map. Near that sacred site with the rock paintings. There's this pile of whitish stones, fifty metres from the track. Easy to find.' He smoothed the paper scrap and June brushed pie crumbs from the middle and studied it.

'I know that place,' she said. 'Special, eh?'

'Better stay home, Aunt. You got that concert.'

'Tidy up that passenger seat. I don't want a whiff of yarndi on my pants.'

'No. I'll be okay. Duckie might drop by later. Bake a cake.' Thommo walked away with a laugh.

She watched him step into the old Toyota. He put his bare foot up on the steering wheel and waved.

The place he'd indicated on his map was sacred, a place which evoked a fourth dimension. Where 'feather feet' spirits from Dreamtime crept around, leaving no marks on the ground. The ochre shelter was a place where ancient spirits spoke to Elders. Even old Aunty Mary witnessed their presence and talked with them in language under black moonless skies. Beings of power traversed the land, creating sites, mountains, waterholes, animals and humans. Old custodians of Country remembered ceremony to protect the site. These beings permeated the walls of reality and perception, but June feared for Thommo, who might not recognise their importance and danger.

One day while she camped near that site, a huge water serpent came – Waarway the Madukka serpent, a being from the night sky of the Milky Way – visited June in a dream. Standing up to strike, showing the power of Karapuli. The pointer stars had men fighting the snake with spears. Its yellow eye pierced her own eye in the night, and she woke in a terrible certainty that the spirits were summoning her. Like the unlocking of gateways with ancient memories behind each one. Like Guthi-Guthi, who placed one foot on Gunderbooka Mountain and one on Mount Grenfell, this creator splitting a mountain to release a water serpent. It was him. The serpent smelt people with a flickering tongue. Smelt her. June knew about her mother's Country, around the east of the Darling River. All her people were descended from either Eaglehawk or Crow. Two matrilineal moieties, two groups. Makwarra and Kilparra.

* * *

A few days passed, and June had a desperate call from Thommo's wife, Duckie. She hadn't been able to locate him. He was missing. He'd never made it home and his phone didn't answer. June felt a surge of panic.

She called the police and told them her cousin was missing, that she knew a location to search and thought something terrible might have happened. She argued and threatened, and eventually they agreed to go with her to look for him.

Hours after her call, the police followed June and her brother William on foot along the track. She staggered on some gravel, but William held her arm tight. She braced herself. They turned a corner near the rock overhang, but there was nothing, no body. Nothing but a heap of sand – no blood, no evidence, just lots of truck- and bike-tyre tracks, and some curious deep gouges in the earth. June stood dumbfounded.

'You certain this is the right track where he was meeting someone?' said Sergeant Blackett. 'Might have been another place?'

'This is it. Someone has taken him. Or worse.'

'You're the investigator. Thought you'd know the answers already.'

William scratched his grey beard and stared with contempt at the town police officers. They seemed dubious about June's reporting. As if she was a hysterical female who knew nothing, instead of a respected Elder.

June sighed, sat down on the stones and held her head. It hurt. She had a deep sense that the young man was dead. She was the only one who knew absolutely where he'd been going and why. Shuddering: with violent fears and also the realisation that she would have to search for him. Alone, if necessary.

She turned to the men. Tired of the superiority of them all, the arrogance of simply having dicks.

Blackett picked his teeth with a stick. He stared at her mouth and snapped a photo on his phone. His hand shook. 'So, you reckon it was a marauding serial killer or something?' He smiled.

'You might be next. What was his MO? A shovel, you reckon? What kind? Shifting shovel with a long handle or a little garden spade? Or a pitchfork, like the devil?'

She imagined a shovel through Blackett's head. See how he liked it.

'You're not stoned on acid or something, June?' he said, looking disgusted as he scratched scabs on his hands. 'Anyway, no body, no crime. We're going back to the station. He's just pissed off from his missus, happens every day. You're reading too many crime books. Too caught up in your TAFE course.'

June couldn't work out why Blackett seemed nervous at the scene. Like he was in shock.

She spotted a print of a boot in the dirt. She pointed to it, but Blackett shook his head.

Soon she overheard him talking to his constable. Blackett clutched his mobile. 'How do you delete a dumb text?' he asked.

'I'll show you, Sarge,' the constable said.

'Nah, just tell me, don't want you reading a private communication.'

June watched the copper, alert with a sort of fear. The boot print looked fresh, so she snapped a picture with her phone. Strange markings imprinted the earth. The police were tramping over it. She couldn't work it out. Where on earth was the lad?

William drove her back to town. Unbroken fields of cotton shone white in the midday sun.

That night in bed, June rubbed her cauterised, dark skin with its raised scars and tried to imagine her deep, fluffy possum-skin cloak. Oh hallelujah, praise the Lord, as she hid from a long conga line of gunjie arseholes, what kind of terrible nightmares would she now endure? Her head bulged. Like a pile of stones waiting for a sacrifice.

CHAPTER 2

31st January

A few days later the family went for a drive to look for Thommo; old Uncle Jack needed a trip out of the house. It was mid-summer, and the heat melted the bitumen. They searched back roads but saw nothing, and were despondent while they drove back to town as night fell.

A siren sounded in the distance. June's young niece Arana looked out the back window: a police vehicle sped after them. It swerved and pulled them over. Arana elbowed her mother. William stopped.

'Shit!' June said, then motioned to everyone in the car to be quiet. Jack pulled his jumper over his head. Sweat and tension poured out of the car upholstery, and they sighed. William turned off the engine as the police wagon pulled up. The air was thick and June's mouth dry as dust. The police officer strolled towards them. A torch in their eyes. Arana seethed.

'The machine has recorded you going over the speed limit,' said Sergeant Blackett.

'We aren't doing anything wrong!' said Arana. 'Drive off, Dad.'

'Oh, it's you, Sergeant,' said William. 'I apologise, it wasn't more than a few kays.'

'We've got a bit of business,' Blackett said. 'That right, fella?' He leant down to the open window.

'Sorry about that mistake,' said William as Sergeant Blackett walked around kicking the wheels.

Arana's iPhone filmed the policeman.

'Put that away,' said June.

Blackett loomed in the mist beside the vehicle and smiled at June. He winked, and she felt ill.

'Looks like a bald tyre,' he said as he came back around. 'Pity, it's a nice old Holden. Collector model. You own it, do you? Show us your licence, sir.'

'I'll see to the tyre as soon as I get into town, officer,' William said, and handed over the licence.

June leant out and grimaced at Blackett. 'We're doing your job and looking for that lost fella. This is harassment. You're only pulling over blackfellas?'

'Quiet, sis, for God's sake,' said William.

'Get off your high horse,' Blackett said.

'They wouldn't let me in the pony club – wrong colour,' she said. 'I hear you're the one known for riding, but not horses. Why aren't you out looking for Thommo?'

'Keep it up, sis,' groaned William. 'Ignore her, officer, she sometimes doesn't know what she's saying.'

'Yeah, I'd shut my mouth, miss,' Blackett said, 'or I might find a bunch of defects on William's car.'

'You call me a defect?' said Jack.

'Why haven't you found that lad yet?' June continued. 'We know he's dead! Rumour has it someone dismembered the young lad's body and he's buried next to the police station. But I suppose you're too busy.' Her chin stuck out.

William smiled. 'How's Mavis? Still working for the church?'

'Flower arranging or something,' said Blackett. 'Your sister there, the big woman investigator, wrote letters to the *Gazette*, stirring up trouble with irrigators.'

'I have rights,' said June, 'I pay rates. Do your job!'

Arana stuck her phone out of the car and clicked a photo.

'Causing trouble with those agitators about water,' said Blackett. 'Hope you've got your Private Investigator Under Supervision Licence? You use unauthorised surveillance techniques. Might have to inspect your qualifications.'

'Oh, you're still here, Sergeant Shithead,' said Arana.

'Just keep your nose out of irrigators' business! Or else.'

'I will risk my life to save our river, I don't see it as a resource for profit,' said June.

'Can we go now?' moaned Arana from the back seat.

'Sergeant, you wear a bit of cotton, do you?' June said. 'You getting a discount?'

'I actually wear a lot of poly-cotton – easier to iron.'

'Gives you BO,' said Arana.

'There's nothing in the back. Nice talking to you, officer,' said William. 'By the way, anyone hand in a shovel? It got pinched from my shed.'

'Nope, no shovel. I could arrest that June … See to that tyre! And you owe me, remember?'

'Get deodorant, they're cheap. And find Thommo!' yelled June as they drove off.

June's sister-in-law Merle scratched her head. 'What's he talking about?'

'Who knows?' said William.

* * *

The night was too hot. June dreamed of the earth after the cutting and chainsawing of gum trees and rainforests. She saw the drying rivers and gasping soil. Then, on the edge of sleep, June tossed in her bed as she saw terrible pictures of Thommo bleeding and reaching out his arms for her, tears running down his dear face. Oh God!

It was a blackout in the town. She reached for matches and a candle. Complete darkness enveloped her. A few minutes passed and a presence appeared in her room. She gazed with wonder as the eerie light danced above her bed. A spirit goonge gathered up bloody particles and flew in the blackness, right there in front of her. It was terrifying. She saw Thommo rising in the air, his face shining with blood, and he smiled, his ghostly hand reaching for her face. She cried out and hid deep in the bed, shaking and cold. So cold.

At dawn she climbed from the sweaty sheets and realised that Thommo was an ecological hero. She might meet him in heaven one day. Poor boy.

June would have to do something to find Thommo. She pictured herself as a brilliant private detective like she had seen on a TV show set in some African town. A brave woman fighting injustice. Yes – she would solve the crime, and her detective agency would be famous, and she would get floods of clients. The black cockatoo Biliir was her totem, and she would fly. It was painted on the window of her office.

* * *

Next morning, the sound of warbling currawongs and galahs burst through the quiet. June's place in Wilga was neat, with pretty lace curtains. Kind white neighbours watered their gardens and nodded to her. The pink fibro walls of June's house looked bright and the pathway from the street shone with white pebbles.

The kitchen had orange-patterned linoleum and 1970s furniture. A metal-cast goanna in ochre colours hung over the front door. This symbol of identity discouraged robbers from the poor end of town.

June walked outside in her dressing-gown to examine her geraniums and roses. She clucked her tongue at their neglected state and pulled grass from the roots.

Old Aunty Mary walked past on her walking stick and called out to June: 'We got lotta trouble here with that river, you gunna help us or what?'

'I will,' said June.

'They all say that. But you do it, girl. We counting on you, find that boy, eh?'

'Love you, Aunty. You got enough tucker at home?'

'I'm all right. The young ones are too educated and now got big money coming in. I got nothing to show except my old house, still got earth floor. We lived on emu and kangaroo rissoles. Eat that goanna too, blue and yellow one. Bush tomatoes, snotty gobbles, eat quandongs raw. I got a whole roo leg in the fridge – I'm gunna mince it up with onions.'

The Murries down the Reserve called June an 'uptown black'. Or a 'coconut'. Not from that country not a Traditional Owner. She didn't care, really; she knew who she was.

CHAPTER 3

3rd February

Bam Bam the bikie pulled up his ancient Indian motorbike outside Duckie's shack. He had Thommo's mobile phone in his hand. The fifty-five-year-old stroked his beard and ran his fingers through his mullet, then checked his face in the side mirror and wiped his white moustache. He sucked in some breath freshener and ambled up to the gate. This shack was in a lonely spot a few hundred metres from the rest of the houses on the old Reserve. It was a decrepit building that farmers had used for storing hay. That hippy Duckie had made it presentable with flowered curtains on the windows. Outside stood a water tank and an outdoor toilet. She cooked on a primus stove and sometimes lit the old wood-fuel stove to heat water or bake a chocolate pudding. It was rustic but homely for the couple and their small son, who slept peacefully in his mother's bed, sucking his thumb.

It was a silent night, and the glint of candles shone through the window. Bam Bam noticed Thommo's clothes drying on the line, and he stood in front of the old jeans. He pressed his face to the cloth, stood back and spat on them. 'That's for taking my woman, you shit-for-brains. I wouldn't give you nothing, or the smell of my fart to cool your soup!'

He padded to the front door of the dilapidated house and scratched on the door. 'You home, Duckie? I found your hubby's phone.'

He heard her breathing near the door. She would be anxious about night-time visits from bikies, even if it was him. She knew him too well.

Duckie opened the door a bit and peered out.

He stood limply at the door with a smile on his face. A kind, generous smile from a bloke who gave presents to sick kids for a charity. He was feeling powerful and in command of the universe. Amphetamines and steroids will do that for you.

'What you doing out here?' she said. 'I don't like visitors after dark, but do I? Thommo will be home soon.'

He smiled and smoothed his grey beard and held out the phone. 'Be friendly, Duck, I just was worried about finding the phone out on a track and I recognised it as Thommo's because it's got a sticker, "Save the Darling". A few locals have that plastered on their phones. You know I was a copper back in Sydney? You can trust me, eh? Can I come in a tick? It's cold out here. I won't hurt you, sweetie.'

The door opened and there was Duckie in her transparent nightie, though she held her dressing-gown closed. He imagined the patch of red pubic hair through the polyester material.

'Don't wake my baby, shush,' she said. 'You can't stay, but – I'm waiting for Thommo.' She wrapped her gown tighter around her body, hesitating to open the door more than a few centimetres.

His bikie boot pressed open the door, and he pushed her back and slithered into the lounge room and kitchen. A fire burnt in the fireplace, a candelabra dripped candle wax on a tablecloth, and incense burnt. He placed the phone on the cloth and patted it. A clock ticked.

'There you go, I'm a good neighbour, eh? You make me a chamomile tea. Got any honey?' He wrung his hands, with their blue tattoos and gold signet and onyx rings. He licked his blubbery lips.

'Where'd you find it? He didn't call me.'

'He lost it.'

'You want that cup of tea, but, or I got a beer.' Duckie moved to the sink to wash a teacup. She grabbed a kitchen knife and placed it on the bench.

'Thought you'd never offer. You've got nice manners and still have nice tits.'

Then he was behind her. She dropped the cup back in the sink. Now he'd found a chance to test out his hard charm on someone soft. Leaning forward, he turned her around to face him, took hold of her arm and pulled her chin up to his face.

She hissed and shoved back. Her face contorted and she spat at him. 'Piss off!'

He rubbed the spittle off his chin, his face red. This was not what he'd expected. He wanted her to recognise that he was a better man than her weakling greenie partner. The world was spinning. He lifted his head up and edged very close to her mouth, staring at the full crimson lips and white teeth. 'Open your mouth for me, sweetheart. Let me taste that pink tongue.'

'Piss off.'

'I got something for you. You remember what.' He breathed in and panted a little.

She stepped backwards and her hand felt behind her for the kitchen knife, but he was quick and slapped the knife to the floor. 'Time to go, mate,' she said. 'I can't have you here. I got little Billy boy sleeping.' The weapon glinted in the candlelight.

She moved, and it was a chance for him to take revenge for her disloyalty. He swung his fist; she couldn't have stopped him. A sudden punch to the face and she was down on the floor whimpering.

He watched her feel for loose teeth. He tore off his leather belt, with jeans around his knees, and tried to grab her hands and unwrap her dressing-gown. 'Should film it on your bloke's phone. A little present for him, eh?'

His eyes bulged, and she grabbed the knife from the floor and held it towards his throat. His eyes burnt into her.

A noise, and he saw the child was awake. He watched the boy standing by the bedroom door holding a plastic *Star Wars* sword.

Duckie stayed quiet. He was king of the world; he could have anything and anyone. But she shook her knife in his face. He laughed. She wasn't going anywhere, but the kid stared.

Next, he pulled up his jeans. He grinned at the boy and rearranged her dressing-gown to cover her nakedness. She cringed.

'There you are, all tidy,' he said. 'Mum's okay, kid.'

The little boy ran to his mum.

'You come near us,' she said, 'and I will kill you, but. I swear I will find a way. Thommo will be after you.' She held her head up and hid the weapon in her pocket. Her hand found her child's head and she stroked him. The red-haired boy ran at the bikie and hit him again and again across the leg with the sword.

'Oh, look out, Skywalker! Hey, he might be my son – how would I know? Eh, Duckie?' He winked at her and pushed the sword away, then smiled and gently nudged the boy towards his mother. 'It's okay, matey ... So, Thommo will come after me, will he? Tell him to bring his own shovel! Limp-wristed hippy soft-dick, wrecking the water pumps. Who does he think he is? Wanted sticks of gelignite and detonators, and thought he'd stroll along to the big cotton farms and blow up their pumps. Boom! Idiot!'

Bam Bam strode out the door. The sound of the motorbike roared as he rode off and the bush was still quiet as if nothing had happened. Some night birds flew overhead, and he imagined her clinging to the child, weeping, then picking up Thommo's phone, but it would be quite dead. He should go back and retrieve it.

CHAPTER 4

6th February, Waitangi Day

After a few days, June had a call from a frantic Duckie. She complained that the police had told her that Thommo was a missing person, and nothing more. They would not help her to find him, despite her telling them about her visitor and the phone.

Duckie pleaded with June to find what had happened to him. 'I can pay you. When I get my pension payment, but.'

'Don't worry about it – no one else does. He's my cousin, I'll find him.' June put the phone down.

Another few days passed, and they realised he really was missing. June had the awful realisation that she would be the only person who might be able to find out the truth. Her teeth chewed at her lips; she leant her forehead against the keyboard of the computer. It was all so exhausting. She couldn't trust the local police to investigate; they had decided Thommo was a bloke who had run off from his missus. The fact that he took no clothes or money – and that his phone had turned up – did not change the police opinion. They buried every piece of evidence. Duckie gave the police his phone. It was wiped of information, useless.

Did Thommo stumble upon someone digging something up? From a crime, maybe? Who was Thommo going to meet? Why was his phone wiped of messages and everything? Maybe he was dead, and the killer didn't mean to kill him but wanted to stop being recognised. That would be manslaughter, and the killer just wanted to protect the identity of a heavy gang of drug dealers?

Then the night-time event. Late and quiet in June's house, he appeared as a ghost at the window and scared her half to death. Bloody face and white hands running down the glass. *Oh, hell.*

June ran from her room into the lounge and called out to William: 'Quick, he's here, he's really dead, that poor goonge is calling for help!'

Possessed is what they became, possessed by his spirit floating above. He wanted to be with them in his dark place. Forever. The morning was not a time to think of forgiveness. They would drift in stars. Curlews cried. They could not count all the dead, calling out to them. 'Give us the dead. We want to possess them.'

The family gathered around her, and they were equally terrified of the supernatural event, but only June had seen him.

'We gotta get some gum leaves tomorrow, burn them and give the house a good smoking. Now we know what happened to our poor fella,' William said, as he shut all the curtains and made tea.

The next day, June asked how she would get people to talk to her. Arana placed sticky yellow notes on the wall of the downtown office. They wrote up a list of suspects in Thommo's disappearance: the politician Mr Warren Wright; the cops, especially Sergeant Blackett; a dealer of drugs in town, name unknown, who they called Ice Man; Gloria Wright, the president of the Wilga Chamber of Commerce, who wanted tourist dollars not river action and was also president of the Regional Women's Association; any of the notorious town bikies such as Bam Bam or

Cowboy; Duckie, the angry partner; or an unknown person from out of town who was just crazy or a Nazi.

June looked at the list and groaned. 'This will be very difficult; no one will want to talk to us. But we can't just sit back and do nothing, Thommo's spirit keeps bothering me.'

* * *

It was early in the morning, but the sun was high and blindingly white. Would it ever rain again? June and Arana walked towards the river in the heat.

The spot on the riverbank held memories for June; she had played with little Thommo in the fresh clean water. He'd laughed and jumped from a rope swing on a coolabah tree right into the river and into her arms. He learnt to speak local language, and she made sure he came to stay with her in the holidays. Thommo was her favourite cousin, and he loved her like she was his mother. His father had run off and his real mum was too young to be competent. Thommo was very bright and wanted to go to university.

'This is Thommo's special spot,' said June, and she showed Arana a picture on her phone of the boy. His smile shone and his cheeky nature glowed as he danced in traditional Aboriginal costume.

'That was at the school dance, years ago. We need to find out what happened to him, eh, Aunt?' Arana sighed.

'If we don't investigate, there'll be no justice. The gunjie don't care.'

Ashes to ashes. June saw the world more crisply than before. It was a revelation, the river nearly gone. Stunted blackberries with tiny shrunken-head berries. No wild orange or quandong to nibble. Rubbish everywhere, thrown over the levee bank. Coke cans, chip packets and pizza boxes. Used condoms, a pair of pink lace undies.

They kept walking. Maybe the river would be behind the next hillock. A wide shining expanse of green, covered in black swans.

June looked for a nice spot under a river red gum to map out a plan. The sky shimmering white, unrelenting. The eucalyptus scent. A cicada started up, deafening them.

A young joey bounced up to its mother and stuck its head in her pouch. The animals looked dehydrated; the murky water was shrinking into puddles. Could the humans fill up a tank for them? But the human drinking water was almost gone too; the bore water tasted like salt and sulphuric acid, pumped from hell. Not enough to give away to the wildlife. Not for sheep, cattle or horses. Dead parrots littered the ground. Thommo called it 'Nature's Armageddon'. The final gasp of the diversity and ecosystem.

Arana flicked on her phone. 'Aunt, I have found some information about Big Cotton on the net. It's what my friend Steve was investigating. I saw their CEO on TV, shiny Armani suit, R. M. Williams riding boots. Government ministers were implicated in doing the deal. Then the drought came, but they all reckoned it was good value for money and said it was for environmental water. Not a drop went to the river. All harvested. Shitheads!'

'We probably won't find out who really committed crimes against our river.'

Arana clenched her fist. 'Remember in the seventies when the press said the Gurindji wouldn't get their land back from Lord Vestey? They never gave up! And we won't either. I told Steve, and he said it all builds a legal case,' sighed Arana.

'You like him a lot, eh?' said June.

'Never felt like this,' said Arana.

'You taking precautions, you know? You can get a contraceptive implant these days,' said June.

'I'm not stupid.'

The Darling River had stopped flowing into the Murray River; it was unbelievable. Climate change and the huge environmental disaster right in front of them.

'This tragedy is man-made,' said June. 'Human beings are killing the life of the rivers. Five years ago, the federal government allocated thirteen billion dollars to save this river basin. It looks a darn sight worse.' She paused. 'We got work chipping cotton in the early days. Over there, look at the crashed river red gums. Skeletons. Green algae blooming. It's like sulphuric acid. Makes you want to give up.'

'That Dreamtime serpent will be crying now,' whispered Arana.

'No government will protect this river system; us ordinary people have to fight. I remember sitting by the river when I was small, and Dad had a sheep he'd bought, and he tied it up and talked gently to it before slaughtering. We cooked it on coals and ate it, dangling our feet in the white, rushing, clean, clear river. We could drink it then.'

'Now it's like goona.'

A movement behind them and they jumped. They were not alone. Someone threw a wet sandbag at them. It whistled past their ears. A stocking full of heavy murder. It missed June's head by a few centimetres. Some person wanted her to stop asking questions about the river and the missing man.

The thrower must have been close. It made her skin crawl as Arana ran over the hillock shouting: 'Who threw that? Come back, you cowards! We'll destroy you!'

The riverbank suddenly seemed sinister with hidden forces, an evil that crept behind the piles of broken trees and cracked mud puddles. The women shuddered and held hands. They wanted to escape to the safety of home.

June had begun investigating in town about Thommo, but no one would talk. Maybe this was a warning? Her heart beat hard, and she tapped her chest to get her breath. She picked up the stocking of sand and examined it. A black stocking like some sexy girl might wear on a date. She pulled Arana back from the levee bank and slid down the grey dust track.

Arana never understood what the older people had gone through. She seemed to think they wanted a soft life, scared to talk up. She didn't know the ancient people who were alive when

June was young – they had remembered Aboriginal heads from decapitated bodies stuck on posts.

***

Later, Merle and June sat inside the Wilga house on the lounge and drank tea. They laughed and looked around at the family photographs and chatted about the town's gossip.

'That Rara going out?' said June. The front door slammed shut.

'Young fella Steve's in town, and look out,' said Merle.

'Trouble there, too smart.'

'Will heard he's a flash mouth, but he works for our river,' said Merle.

Merle got up and made another pot of tea, then poured June a cup and stroked her hair.

'Picking up posters tomorrow; don't know how we'll pay for them,' said June.

'Oh well, have another Monte?'

'I'm getting fat,' said June.

'Nah, have two. I am.'

CHAPTER 5

14th February, Saint Valentine's Day

The postman left a Valentine's Day card in her office letterbox. It was from a secret admirer. The glitter card had violets and hearts all over it. It also had a gun drawn, aiming at a target. June threw it into the recycling bin. She had more pressing problems.

The local Wilga politician Mr Warren Wright was on June's suspect list. She had drawn up the names with motives next to them. Wright was up to his neck in corruption about water rights. Everyone knew it. He hated river activists.

At lunchtime, June walked straight into Mr Wright's office downtown, making herself at home on the leatherette couch. She noticed it was stained and hoped it wasn't Mr Wright's semen.

He looked at her with contempt.

His pretty sixteen-year-old secretary rushed into the office. 'Sorry, sir, she just barged in, no appointment.'

Mr Wright waved her away. 'That's all right, sweetie, Miss Private Investigator here has some business. I'll take care of it. Don't worry your pretty head.'

'We need to talk about Thommo!' June smiled and adjusted the cushion under her head.

His mouth dropped open, then turned to a snarl.

She saw him, a wankasaurus, a big swinging dick with dandruff on his shoulders.

'The Shire Council has appropriate protocols for talking to Indigenous stakeholders,' he said. 'You'll find them on our website under the Reconciliation Action Plan. Now, I'm a busy man, so ...'

'I'm not here to listen to your stupid protocols. I want to ask whether you know Thommo, and how you objected to the actions of the Save the River Group.'

'There are formal complaint channels.'

June grimaced, and smoothed her dress down to hide her knees. 'Thommo?'

'Who cares? Bunch of radicals, red flaggers, commies and ratbags, the lotta you.'

'We care about the river.'

'There are appropriate channels to voice these concerns, Miss. The Murray–Darling Authority has an Aboriginal member; ask him. I hear you have a very special relationship.'

'Tell me what you know about my cousin's disappearance, or we might just contact the media.' June sat up and leant towards him.

'About what? You threatening me? Think you have some dirt? I'll not be intimidated. Who do you think you are? Just an effing upstart old gin.'

'We have recordings,' June lied.

'What of? Give us a squiz! You've got nothing, the police are investigating. It's a matter of obeying the law. The constabulary will work it out all in good time ... Who cares about a junkie anyway? Off you go. Or I'll have you arrested for trespassing. Just piss off!'

June pressed herself against his desk. 'You won't win this one!'

'You reckon? Give me a break!'

* * *

The day became very hot. Flocks of grey emus strode past just on the edge of town, and white cockatoos soared in the blue sky. Past and present seemed to collide; time stopped; no one said much; it felt so hot.

William, Merle and Arana sat in June's office eating ham and pickle sandwiches. Arana's book engrossed her; she never stopped studying except to text that boyfriend. But June was uneasy, as if along that dusty road something terrible was coming closer.

They were determined to continue their activism. They collected the Darling River Environmental Action Group posters from the local printer. Merle had boiled up flour and water glue, and they stuck up posters for the town meeting to save the river on the street's electric poles. Before they finished, Arana observed that someone had ripped the posters in half outside the town hall.

'What shitheads,' said William. 'If I catch them at it, I'll throttle them.' He groaned and nodded at June, and they drove into the main street of Wilga. 'You own that bit of property in town, it becomes part of you. It's all right that you run a new business too. Really good, sis. This land is where our grandfather lived and died. It's part of our hearts. Just being near that river kept us alive as kids. You have this tiny piece of our ancestors' land, just that quarter acre makes us all seem bigger. Plant a tree in the backyard and build a shed and nobody can take that away.'

'Except if I miss a mortgage payment,' said June. She stirred the glue pot and glued a poster on a light post on the outskirts of town. People in old four-wheel drives stopped to read it. Nearby, a young man was ripping a poster to pieces with a bowie knife.

Another man in a cowboy hat dropped his shorts to give them a brown-eye.

Arana leapt up and ran towards the young men. 'I'll glue a poster to your stupid face next time!' she shouted as they drove off laughing and revving their vehicle.

They hooted. 'Fuckin' Blacks! Stick your environmental sign up your arse!'

'Nice! Your mothers know you're out?' shouted June.

William threw a rock after them. 'Shit-stirrers!' he bellowed, then collapsed holding his chest, great gulps of air heaving from him. June knelt beside him, undid his collar and held his head in her lap.

'Just asthma,' he said. 'Help me up and I'll find me puffer.' He coughed. 'Next time I'll have a stick of gelignite in my car.'

'Come on, Dad, let's go home,' said Arana. 'This river business is killing you.'

* * *

The Darling River manifested as a Dreaming track, a place where old man Crow chased the spirit women across the red earth. The seven sisters from the stars had stopped and flung goona at old Crow. June pictured the winding path of stories joining the river to the Murray and on to the sea in South Australia. Deep and unending, pulling gently on their hearts.

Next day the hot sun came up, and June squinted in the pinkish light turning yellow in the dawn. She got out of bed and walked across town to the river. She probably looked a bit of a madwoman, so she scraped her hair up into a ponytail. The pink and orange sky took up half the world. So much unending light. A real *Assumption of the Virgin* sky. Where was poor Thommo?

The local farmers complained of drought, of the lack of rain and the banks not lending money. It was true. Probably alcoholics, the lot of them. The farmers suffered from depression; they said Australia rode on the sheep's back to prosperity. June said it was more like the Blacks' back. Aboriginal people worked for nothing, just rations of meat, fat, tea, sugar and flour.

All the topsoil blew away in the wind because they cut down the trees, turning the sky red with dust. Here, nothing but crows squawking. They wanted handouts from the government, relief payments, and the dole. And they sent their sons to posh city boarding schools on scholarships. Well, she reckoned they should give up farming and go back to England. Have tea with Prince Andrew. And they might ask for the warrior Pemulwuy's head to be given back while they were there.

On the way to her office, June made a detour to the Regional Women's Association to introduce herself as a private detective and find out any information about the crime. And try their cakes. She wished she had some sort of badge. To get a bit of respect. She wondered if anyone would know about what happened to Thommo. The shopfront had happy pictures of women with cakes. In the window sat a pair of nice, crocheted pink booties.

A bell tinkled as she sat down at the morning tea table and grinned at the sunburnt women welcoming her. The women were obviously from the landed gentry, talking about pony club and how tough life was in the drought and the cost of sending little Marigold to a good school, possibly Frensham.

June looked around and there she was again: Queen Elizabeth on the wall in a golden crown with jewels stolen from India, and a nice blue frock, a sceptre in her hand.

Sitting on an embroidered cushion, June was handed a plate with a scone, jam and cream. She breathed out.

A few people were mortified because June was in the room, so she mentioned her new role as investigator. Everyone nodded. She picked up her bag to leave and an elderly lady spoke directly to her: 'So nice to have a new member with us.'

June smiled and purchased some pink iced cakes with silver sprinkles for Jack and carefully placed them in her bag. The aged lady took a whole plate of lamingtons and gave them to her. June murmured, 'No, thank you.'

'For the kiddies on the Reserve. We made them fresh, you take them down there. I have never been, you see. Wouldn't want to interfere in their business, it wouldn't be polite.'

Meanwhile, another woman held the door open and tapped her foot.

'Thank you. I wonder if anyone has information about the disappearance of Thommo?' June said and everyone stared. 'Would anyone know about that young man?'

'It's a real mystery,' said a woman in a print dress. 'Except we heard he was in bed with bikies and drugs, but everyone says that goes on these days. They manufacture *ice*, whatever that is . . .'

'No, he was a good lad,' said June.

'I work at the hospital; I've seen that missing man and his wife a few times in Emergency. He came in last year as if he had been hit with an implement. I thought his wife might've done it. She was on something.'

The president, Gloria Wright, leant towards June. 'You know, your cousin once worked for the Department of Roads and knew all about detonators, storing explosives and ammonium nitrate fuel oil. Rumour said he tried to buy a load of it. What for, I wouldn't like to guess.'

Now June sweated and worried, and knew it was true, that her Thommo, the river activist, wanted to blow up Big Cotton's irrigation water pump to try and save the river. Trying to be a hero and it had got him killed. Somehow she needed to find who was selling the explosives, who had met him that day. By finding evidence she could go to the police in the nearby big town.

'Who might be able to supply that stuff?' she asked.

'Anyone who worked in mining, or ex-army, I suppose,' said the president.

'I'll leave some flyers for the meeting to save the Darling,' said June.

June looked at the face with lines crisscrossing its freckled skin. She pictured Black and white holding hands to save the river.

'We are joining with other groups to force a Royal Commission to attend to the state of our river and the terrible taste of the bore water,' said the president. 'It's full of salt and sulphides. What will we feed our children when all the water dries up from the cotton farms? We can all care for this land. Perhaps we can work together?'

'Come to our meeting, I can leave you a poster for the window,' said June.

'No thanks, probably full of drunks,' said a member.

June shut the screen door and walked away, but a woman followed, reaching for her arm.

'I realise a number of white people care, but we don't need caring for,' said June. 'We are not kids – we need respect and justice and employment and some of our land back. The council should start by flying the Aboriginal flag outside its chambers. Why don't you ask them about that? And if anyone knows what happened to young Thommo, please let us know.'

'I will.' The woman put out her arms, and they embraced. 'I can get you some names of the Lightning Ridge miners who have access to that stuff. Give me a poster, I'm sick of looking at a dead river.'

June handed her a rolled-up poster, then crossed the street towards her office door.

Duckie sat outside weeping.

'Come in, girl, tears won't help.' June patted her hand. 'I know you're worried, grieving for him, but I'm working on it. I just keep getting blocked on information about him.'

'He's buried out there all by himself. So alone.'

CHAPTER 6

16th February

June was aware that a shadowy figure lurked near the back window of her office. Sinister and quiet. June saw him cross the road near Fay Pascolini's hairdressing shop. His cap flashed along the corner of the dirty glass. Who the hell?

The sultry morning had begun badly, with a broken lock on the door and signs of someone mucking up her filing cabinet. No chance that police would arrive to investigate a break-in. She felt his breath along the window, close enough to listen in to her conversations. Not that anything interesting had happened, just a whining drunk searching for his lost wallet. But now, whoever the person was, he was scaring her.

June picked up her boondi stick and laid it on the desk. She stroked the mulga wood. Jack had carved it years ago. The stick was gripped in her fist. She would show them. Stepping quickly out the front door, she charged around the shops to enter the back lane. The office had its bins and recycling out the back, and hardly anyone went there except on bin night. But the man had moved on or jumped in a bin. She checked inside: just papers and coffee cups. Her breath was laboured, sweat trickled, and she wished she had not drunk two coffees. It gave her the jitters.

No one was hiding down the empty lane. She appeared nonchalant and whistled a tune; ridiculous, really. The main street had its usual mid-morning foot traffic of men in bush hats walking into Elders to buy a bag of nails or sneaking a smoke at the bus stop. Their wives carried bags of groceries, and children clung to sunburnt legs.

June stood around a corner and watched to see if a suspicious man was watching her office. Maybe he was waiting for the street to be empty at lunchtime, and then he would strike.

Her nerves were raw. Some farming men flicked their eyes towards her, but no one seemed to be spying. What did the man want? Or was he just taking a jillawa in the laneway, and was innocent?

The boondi stick seemed silly; she hid it in the folds of her skirt. Perhaps people would think she was playing clapsticks. June looked in the display window of the hairdresser's next door. In big letters the window read 'Beaux Chevaux' – meaning 'beautiful horses'. The owner had asked the window painter to write the French for 'beautiful hair'.

Then she saw the reflection of the baseball cap, the dark handsome face and wide eyes looking back at her. Who was it? The shadow of him behind her approaching fast – the hide of him, in broad daylight! She swung around with her boondi stick raised, but his hand shot out.

'Aunt! It's me, Steve.' He gently touched the stick, and she lowered her arm. 'Arana's friend.' June shook her head and motioned for him to go to the back door of her office. He peered through the mesh panel at the back. June unlocked the door and he slipped into the gloomy room.

He was very dark with a shining face and beautiful eyes. She realised that her niece was smitten for good reason. He waited politely by her desk. He held his baseball cap in his hands and twirled it with nervous gestures. Meeting his girlfriend's aunt was a big test, and he looked eager to please.

'Gidday, Aunt. Thought I'd better come here first before Arana gets me around to meet her father.'

'You scared me by looking in the back window. Nearly belted you. Better take a seat and tell me about your mob. You look like a nice young fella.'

'I'm originally from Northern Territory, I was born up there.' His eyes were lowered. 'Mum's white, from Queensland, so I was brought up mostly by her. She worked hard as a lecturer at the uni to support me. But I went away to boarding school a long time ago and haven't been back up north. I have to work to support myself, so I just talk to my family on the phone; they're okay.'

The world seemed more incredible with a young man like this coming to her family. June imagined Bangarra dancers on stage with a lead like Gulpilil in their midst. She had been mesmerised by traditional ceremonial songs, and loved attending corroborees along their river. Steve's presence brought a wondrous sense of possibility. The connection to spirit and ochred ceremony called bungal djama. She had read about it all.

'Proud of you, I bet?'

'Guess so. This is all right, eh? Your office looks real professional. You got some law books over there – I might have a look sometime, if that's okay?'

'I have a qualification in responsibility for factual inquiry work, covert surveillance operations: not quite what you're studying but not bad for a dubai who finished school at fifteen.'

'Sounds like a private investigator?'

'I'm learning. Arana tell you about the young fella who was probably killed?'

'Yeah, she's got some ideas about what happened. It's real cruel, that case. Want some help?' Steve smiled and his grin was like a beam of light.

He wandered around the office and bent down to read the sticky notes on the board. He took out his phone to take a picture. 'Is it okay if I take a photo? I can study the suspects for you and ask a few people I know in the police in Sydney. I have some mates who are detectives.'

'Don't tell Uncle Jack,' said June.

'My true reason for coming is to take Arana to the ball. That's the way to a young woman's heart, eh? I've never been to a ball.'

'She's dying to wear that dress.'

'I've written an article or two about climate change and the disaster of the Darling River. There're a lot of Indigenous activists in the city, and we work alongside some major climate strategy services. We have articles in major online news feeds. The times are changing; we can help you up here.'

'You got some smart ideas, that's for sure.'

'And we have a TV crew coming up to do a story about water thieves, so I hope you can get involved too.'

The office phone rang and June picked it up. She listened and then hung up without speaking.

Steve looked worried. 'Something wrong, Aunt?'

'A shitty threat, as usual. They ring up and say rude stuff; we're used to it around here. Racism is alive and well in Wilga. We won't be bowed down; we want to save our Country.'

'I'm sorry an Elder has to put up with that rubbish.'

'Stick around and you can meet some utter pigs, like the local drug-dealing bikies. Make your hair curl.'

The front door opened and in walked Arana. Steve laughed and beckoned to her and she leapt into his arms. He picked her up to swing her around the office.

'You never said you were already here!' Arana grinned.

'Surprised you, then? I wanted to meet the local private investigator and give her a hand to solve some crimes. Want to help out?'

'She's got HSC,' June said. 'We have some evidence and suspects, but local gunjie are hopeless. They don't care about us Blacks. And we must try to keep our heads down – a lot of people don't want me to investigate.'

'Why's that?'

'A lot of townspeople make their living from cotton. You see, it used to be all rural, sheep mostly,' June said. 'My uncles were shearers once, now they're working for the dole. Our people had a lot of pride in their ability to ride horse and muster cattle, too. Even if after that 1967 referendum they had their names removed from the station registers because the government forced the owners to give equal wages. They got nothing but flour and a bit of meat, basic rations, you know. Used to be a lot of work on the farms; all gone now. It's cotton and security firms, but they bloody employ ex-bikies. It's ridiculous – they're thugs.'

'Shops too, Aunty, some of my friends from school got jobs because of cotton. While people live on the profits we can't make progress on saving the river. But what's the point of having a nice job and new clothes if you have no water to drink? Or the environment is finished up, eh?' Arana sighed.

'Reckon that young fella got kidnapped or killed because he knew something about illegal water extraction?' Steve wrote notes on his phone. 'What do you reckon, Arana?'

'I don't want you involved in this investigation, dort,' June said. 'That's final. Merle and me are driving to Wilcannia tomorrow. So behave yourself with this young fella.' She slammed some papers on her desk, then smiled and gave then both a hug. She packed up her laptop and closed the windows and door. They all left the office.

CHAPTER 7

18th February

'No river, no fish, no life' read the notice on the stone wall on the edge of Wilcannia. It was a hot day when birds fell from the eternal blue sky. Flies clustered around eyes and got stuck in your red cordial. This was a town that bikies like Bam Bam liked to terrorise. The dusty town looked downtrodden and poor: iron bars over the windows, a few bedraggled shops. Children of all colours rode bikes down the main road and Murri women with babies on their knees ate chips and gravy outside the petrol station. The sandstone police building stood out in the street; red dust clung to the dry trees and cracked bitumen.

June and Merle were in town to hand out flyers, and they went to the Aboriginal school for a talk about climate change and the river. At the after-school meeting for adults, June acted like undercover police and made some not-so-subtle inquiries about bikies, but no new information was shared with them except a curious comment about the arrival of 'the Ice Man' in town. June wrote notes on her phone and kept up a light banter. She didn't want to frighten the nice teachers.

The classroom was picture-perfect, with flowers in the garden and bore water to keep them alive. All the classrooms shone with pictures of Aboriginal people and language lists.

June had seen the bad press the little town received about violence and drinking and domestic abuse. Rumours tore through the town about Thommo. They were scared.

She looked around the classroom and the people were well dressed; all she could see were Murri people with a keen desire to learn the future of the river. They had been denied the truth for so many years, with historical stories of valiant English explorers and white pioneering spirit shoved down their throats. Now the families were furious. The equivalent of five Sydney Harbour's full of water were siphoned off by the big irrigators, most of whom were working for overseas owners. There was the growers' ongoing sabotage of water-pumping meters, which had been put in place to measure the flow into their dams.

Merle pointed out a sign on the wall drawn by children, a blue curving river. The writing said: 'Oh, the Darling River. We are the people of the river.'

'Elders are giving up out of sadness.' A teacher spoke. 'We want the water to flow again, as it gives hope and life and brings us together.'

June listened and nodded, but out of the corner of her eye she saw two white men with beards sitting up the back. Cowboy and Bam Bam. What on earth were they doing at such a meeting? She felt a surge of fear. They looked out of place, with leather jackets and the blue tattoos on their hefty forearms. They wore Gobbler vests. Bam Bam's beard was plaited into a rat's tail with a shiny bead on the tip. Cowboy wore tattoos and little white shorts and a beaten leather jacket.

There were stories about the bikies camping near a dope crop on the outskirts of the national park. Cowboy guffawed and then shoved his fist in his mouth to stop laughing. Stoned, probably.

The room was silent; June sighed. 'Maybe you're at the wrong meeting, mister, but I'd like to talk to you afterwards.'

One man rose up and winked at June, then he motioned to the bikies to leave. He paused near them and waited, and they walked out. Bundi was a local Elder from Dubbo who threw his weight around western communities and worked part-time as a university lecturer back in Sydney. Tall, handsome with nifty shirts and Italian pointy shoes. He looked angry about outsiders interfering with procedure. He was associated with the Murray–Darling Basin Authority. There would be consequences for people who mucked up his work.

'Payments to irrigators are legitimate!' said Bundi. 'We Murries have worked alongside the farming lobby to ensure our legal rights.'

Later, there was a community barbeque, with Merle and June talking to everyone who wanted to fight to save the river. It stirred people up, and the pictures of the protest on the dried-up riverbeds shone from noticeboards. June admired the waving mass of red, yellow and black Aboriginal flags.

June and Merle approached Bam Bam and his giggling sidekick. They beckoned him to stand under a tree away from the rest. Bundi arrived next to her and touched her shoulder, but she shrugged his hand off.

'I hear you're asking inappropriate questions about my brothers.' Bam Bam smirked. 'You should watch it!'

'Where were you when my cuz Thommo went missing?'

'Hey, June, that's enough, eh?' said Bundi.

'Nowhere near him, don't know him, don't care. I heard he was muscling in on some important growers, big blokes, deserved all he got. Try knocking on Mr Wright's door.' Cowboy blew up his chest and pushed towards June, but she had Merle and Bundi beside her. She gave the bikie a little shove.

Cowboy looked at the dark hand with the silver ring on his chest and he laughed. He coughed on the hand. 'Don't mind me covid,' he said.

The intruders moved over to the Elders' group, where Bam Bam apologised for his friend and the two bikies walked out of the barbeque, clutching sausage sandwiches and wiping tomato sauce from their fingers.

* * *

Late in the afternoon the sun was setting golden through Wilcannia red gum trees. William made a camp on the dry riverbed and lit a fire to cook dinner. He laid out their foam beds on a blue canvas, wrapping them up in blankets because the night would be freezing. He smoked his last cigarette and checked his watch. Those women were supposed to walk back from town to the tent.

He heard the sinister roar of a car revving nearby. He jumped up and checked if his hunting rifle was still in the boot of his car. He picked up a box of ammunition, then thought better of it. The noise might have just come from Murri lads having a grog.

Then, a distant cry, like an animal in distress. It was a woman crying out for help!

William peered through the shadowy trees and called out, 'Who's there? Are you all right? You need help?'

The sound of a ute burning up the dirt, and calls from young men echoed. 'Scum! Watch out, you shit-faced Blacks!' someone shouted in the dark. Cars revved and tyres performed burnouts in the riverbed. More frightening: a blast of a shotgun. The light of the shot lit up the ghostly trees, like a small fragment of burning hell. William ran towards the sounds, panic setting in. What if the scream came from Merle?

William puffed as he held his heaving chest. He should have got in his car – on foot he felt out of condition, an old bloke. The yelling grew louder, and he realised someone was under attack. He would rip the youths limb from limb.

He rounded a corner and there was a scene with the women from his family, surrounded by white utes full of gubbah white fellas yelling at them.

'Help, help! We'll call the police!' yelled June.

The vehicles had roo-shooting searchlights fitted on their roofs and roll bars. One man stood on the back of a car with a shotgun. Their cowboy hats marked them out as cow cockies' sons, while the newness of the utilities shouted wealth and power.

The young fellas turned to point out William lumbering towards them. He was shouting: 'You leave those women alone! I'll bloody well kill the lot of you!'

The cars backed up, and in the lights William saw Merle cringing and holding her blood-streaked head.

Beside her, June had a lump of wood, and she rushed towards the closest ute, hurling abuse: 'I know your parents! I'll come to get you! The gunjie will charge you if it's the last thing I do!' A young man jumped off the vehicle and threw a rock towards William and then a can of beer as June caught up with the ute and cracked a headlight with her stick. June turned on the man, who wore a white Akubra hat.

He ambled towards her with his fist raised. 'Bugger off, Black stirrer, we don't want you coming up here with your stupid protest meetings. That young bloke went missing ... Well, you're next, Detective Cluso!' He leapt back in his car and revved the engine.

June stood still, controlled and courageous before the vehicle.

'We eat greenies for dinner!' shouted the youth with the shotgun. 'Watch out you don't end up like Thommo!'

William picked up the can of beer and smashed it towards the man. He missed.

'Let's go!' yelled the driver.

The vehicles roared back up the dirt bank, doing wheelies, the searchlights like eyes on the ghostly trees. The trucks threw dust over William and the women; they were choking.

June helped Merle along the track and washed her head with a billycan of water.

'I can hardly feel it,' said Merle. 'Just kids letting off steam.'

June wiped antiseptic ointment on the scratch and pressed the band-aid over it. She kissed Merle's forehead, then said, 'They had a gun! Maybe they killed Thommo?'

'I'm packing up,' said William. 'I'll call the gunjie in Wilga; at least they know us; we can make a report. I reckon they heard you asking questions around town. Wanted to shut up your investigation.'

'Thought they were going to shoot us,' said June.

'I never felt comfortable in Barkindji country, too many goonge here.'

'Glad *you* didn't shoot *them*,' said Merle.

'I would have, but I left the gun in the car,' sighed William. 'I don't understand why cow cocky sons would not support sustaining the river. They have no water.'

They climbed into the Holden and drove through dust up the riverbed.

'We've got to get home.' Merle smiled. 'Arana might get up to something with her fella. All alone in her bedroom.'

'Aww, don't want to know that!' said William.

'I got her some condoms. Black Sax brand,' said June.

'Shut up, sis.'

***

June thought Thommo was dead, that it was an unsolved crime and she needed to work it out alone. The police were unwilling to help. June had written out the details of evidence she had gathered. Her hopes of a proper police investigation were fading.

She had written many letters to high-up officials and made countless phone calls. One day she sat for hours on the seat in Blackett's office with a placard that read: INVESTIGATE THOMMO'S DEATH NOW!

Perhaps Thommo was killed because he had big plans, like a terrorist, making a big splash on TV – but why kill him? What were the conspirators afraid of? Who exactly were they? Surely the coppers would've arrested Thommo if someone dobbed on him. Maybe the supplier of explosives was linked to someone high up? Like a set-up to discredit the Murri protesters?

She grasped at conclusions, wanted to investigate but not draw too much attention to herself or her family. Someone must know the truth. Who had what to gain by his death? Could it have been Duckie? She pushed the thought away. Too cruel. She knew that a big corporation was trying to buy Aboriginal land. That their Country had been won back through twenty years of struggle by the Wilga Aboriginal Land Council. Now, the international consortium wanted to get it and start a huge almond farm and suck out what was left of any rivers. She thought it smelt of Bundi's involvement, but couldn't bear to consider him a traitor.

Then dear Thommo had opposed the motion to sell at the Land Council Board meeting, with shouting and doors slammed. A real dust-up. He had called out the disgrace: yuppie almonds growing on Aboriginal land set aside for the children of Wilga. Bundi had just sat there with the papers for signing in his hands.

June asked herself if this sacred land was so treasured for its economic value that someone in the way would get killed. Who would benefit most from the sale? Surely the Aboriginal community owned it all and most wanted it protected. Who organised the murder? Who got paid? Crimes like that happened all the time in Sydney. Was she in danger too? Perhaps she should give up trying to be an investigator. She shuddered, turned on the TV and tried to forget.

CHAPTER 8

19th February

The next day, June spent the hot afternoon writing on yellow bits of paper and arranging them on a board to investigate the murder of Thommo. His young wife was on the phone night and day, begging for answers. She said she regretted giving police his old phone.

Duckie wondered aloud if the bike gang was involved. June explained carefully how Duckie must keep her opinions to herself because those men were extremely dangerous. But Duckie could not bring herself to accept that her man was killed over the pathetic amount of dope he sold. It was barely enough to pay their rent.

Thommo had been travelling on and off to the Adani blockade. He had recently recorded illegal pumping by local irrigators, so she felt it was all mixed up; no one would tell her the truth. June couldn't tell her anything new. In fact, she kept the fragments of evidence from everyone.

'We need to find out who drove his Toyota to the dump,' June said. 'They'd get some fibres from a flanny shirt left behind, but it's probably Thommo's.'

'Thommo was a trained bulldozer operator and stuff; he had a job once, you know Department of Roads,' Duckie explained. 'I remember he helped build the new highway bypass a few years ago outside of town. Everyone had said what a good job he done, but, except Aunty Mary, who had put a curse on him or pointed a bone, but I don't believe in oogie boogie magic.' She burst into tears.

June hugged her and handed her a tissue.

'He's been coming to knock on the window up home!' Duckie's eyes grew wide, and she chewed her nails. 'I know he's dead. And because the nights get real cold, but, up near the Reserve, I light a fire and leave the light on, but I'm so scared.'

'Of bikies?'

'Yeah, the fuckers. I want to get a gun and blast the lot of them.' Duckie clenched her fists. 'Vengeance, you know, like in Netflix, revenge. It'd be *The Shining* – all blood, guts and stuff but. You see, at night, but. He's crying at the glass. Shit myself.'

'Who? Bam Bam?' June looked horrified.

'Nah, Thommy. I think I won't be letting that ex-hubby arsehole in me house, but. I'm moving to town, Women's Refuge first off. I mean, what's the point of going and telling on him. No one cares, but, there'll be no come back, they'll just think *she's a slag and asked for it.*' Duckie hung her head.

'Talk, girl, talk!' said June.

'That bikie hit me.'

'Sure. Come into my bedroom,' said June. She put her arm around Duckie and escorted her out of the room.

Ten minutes later, June showed a crying Duckie to the door and pressed some tissues into her hand. 'We'll find out what happened. You can depend on us. Your child deserves better. But you should've gone to the Women's Health Centre – they can help. It's trauma.'

Duckie wiped her nose. 'No, keep my secret, please! I'm frightened.'

***

The next morning, June was ready to take revenge on Bam Bam. 'It's time to get the Murri women together and teach that bikie coward a lesson. I can't stand listening to Duckie's pain. Gunjie have done nothing. There are consequences. Violence against women is not tolerated anymore.'

'Yeah? Tell that to the fairies,' said Merle. Still, I'm in. I've got Jack's boondi stick.'

Jack looked alarmed, but was focused on the footy game and couldn't really hear them.

William was out, thankfully.

'It's dangerous,' June said, 'but perhaps the shame of a beating by dubais will keep Bam Bam quiet. If there's enough of us, we can catch him at his caravan and beat him into begging forgiveness.'

Late in the afternoon, the phone had been running hot, and as many as eight fearless Murri women arranged to meet up outside town. They were all bringing some kind of man-beater stick. They were tired of domestic violence and men getting away with it, and even burning a woman alive in front of her children. This man was about to be beaten up by angry black women.

The clock ticked, and at four o'clock three other cars pulled up outside June's house. One car was from the Aboriginal Land Council. They all wore black hoodies as if they were off to some midnight robbery.

June climbed into William's car and hooted the horn. They all drove off, fast.

'Bam Bam will be home because I've heard he has a nana nap at four,' June said.

'He'll wish he was at home with his nana when we finish with him,' said Merle.

June pulled up in the car and talked to the other women. 'I know some of you think that because Duckie is white, we should leave it alone. But she is a cousin, and we owe it to her hubby to take revenge on skunks like Bam Bam.'

Respect for the Elders was important, and June needed to ask permission down the Reserve to carry out such a punishment. The shack was hot, and the tin gate rumbled as June walked in. Old Mary was sitting at her table, drinking tea. She looked up with glaucoma eyes.

June explained the situation.

'Let it go,' said Aunty Mary.

June stirred her cup.

Aunty Mary continued: 'Something will show us that truth, like a dream or vision. You'll wake up and know who killed that lad.'

June smoothed the tablecloth.

'I don't want to be bothered by all that rubbish,' Mary said. 'I can't talk, I might have a stroke or something. That Thommo had no kuyan shame, lot of people didn't like what he did.'

June nodded at Aunty Mary and went back to the cars where the other women waited.

'She knows something,' said June. 'She's cagey, but she has second sight and talks to goonge. She won't talk about this stuff, thinks we're stupid to dig around. Her generation were beaten if they even spoke their language. She's nervous. It's understandable.'

The women drove down the bush track to Eagle Nest Point, where Bam Bam kept his caravan. Galahs squawked and an eaglehawk hovered overhead. It was an old graveyard of trucks, utes and motorbikes rusting among tussocks.

'That kilaa and maliyan there,' June whispered.

'Minya?' Merle said, shaking with nerves. 'What? Let's not go in there, Bam Bam might shoot us!'

'We need to show that warraaykuuy he can't get away with threatening women. It's retribution.'

'You're too hot-headed.'

Leaning down to examine some boot prints, June took a photo. Surely it was the same boot from the murder site.

She crept up to the caravan. The backyard had piles of motorbike parts and a concrete bunker that June thought held dead bodies, but looked like a sewage tank. The block of land belonged to Cowboy. June used to see him walking down to the river before it dried up. He took a bath with a cake of Lifebuoy soap, but it was Bam Bam she was looking for. He was the stinker.

June felt an anxiety attack coming on; her chest was tight and it was hard to breathe. The Indian motorcycle was parked in red dirt, another new shiny bike was behind it. Funny how bikies always had a mate, like they were afraid to go for a jillawa alone. June hoped this time he was by himself. Two men might be too hard to take on. She tiptoed alongside the caravan and a rustle in the grass told her some snakes lived close by. She looked back at the vehicles behind and the other women waved with big grins. Too late to back out.

Merle saw the caravan with peeling paint and red spatters. It was foreboding. 'Let's not go in, I want to go home!' She whispered and clutched June's arm, then grabbed June's necklace of painted gumnuts and it broke. 'Oh shit, please, sis, don't go near his van! Look at that chopping block.'

They noticed a bloody axe stuck in the wooden block. Red guts and fragments of flesh hung off the implement. Flies buzzed. A dark metal statue of Satan with goat's legs stood like a shrine outside the door. Silence. Then June motioned to the other cars. She was going in. She smoothed the waddy stick with her fingers. Merle took hold of June's shirt and pointed to the black flag flapping inside the window. A swastika.

'Christ, Nazis and Satan worship.' June grimaced, and Merle froze. 'You stay here if you want?'

'No, I'm coming.'

The other eight women crept up around June with sticks and boomerangs held high.

June motioned to the gang to be quiet. Full of pent-up rage, she stood up on a box, her solar plexus thumping, and looked in the dirty window. He was asleep. His great stomach heaved in a grey t-shirt and his grey beard fluttered on his chest. She gave a thumbs-up and the women poured towards the steps. They opened the door. June saw him wake and heave himself up. He grabbed his machete. Before he could speak, waddy sticks were flaying the back of his body. He didn't know what hit him. Every woman hit his crouching body. He cried out in pain and lashed out, but the machete had been beaten from his hand.

'What the hell? Stop it, you Black bitches!' He smacked at the sticks, but he was a giant attacked by small beasts, and he could not get a single hit against them. He cowered on the ground and June stepped up. His face was red with pain and humiliation.

'Don't you ever, ever try to rape a woman again or we will cut of your tiny booble and throw it to the dingoes,' she said. 'Our sovereign country, our law. You got it?'

'I didn't. You got it wrong, Aunty. I respect your culture and I wouldn't touch a Black cunt.'

'She's a white girl. We know it was you! Go near Duckie and you're gone.'

'She wanted it; she would have enjoyed it! Crap to you!' sneered Bam Bam.

'Did you kill Thommo?' Merle asked.

'Nah! Some fucker did. He took my woman, and he was shit-for-brains, wrecking pumps that belong to hard-working farmers. I wouldn't give him money for dope, or the smell of my fart to cool his soup! He was just another hippy stopping progress.'

'Go back to Transylvania!' yelled one of the women.

June kicked at him but missed. She walked away with Merle clinging to her hand, and the women took off in their cars. June had to stop her ears to their yelping and cheering. Her breath was convulsive. She pictured Duckie smiling when they told her about what had happened. June felt elated and didn't care what consequences might come. They were powerful black women, standing shoulder to shoulder together, from invasion until now. Nothing would break them.

On the way home, June spotted smoky mist from cultural burning going on out bush. She imagined herself with her grandmother, walking through the eucalypt trees and setting small fires in the grasslands. Her granny would chase after ngurruy, an emu, if it ran in front of them and bang it on the back with her waddy. The same weapon June had used on Bam Bam. But at least you could eat an emu. That bikie was useless.

***

The next day, June walked into the lounge room. The television was blasting, and Uncle Jack sat glued. He leapt up when he heard a motorbike outside. She froze. This might be the revenge the bikies wanted. Jack looked out the window, and it was the bikie Cowboy doing smoking wheelies. June crouched behind the curtain to watch. She watched him tear off down the road, leaving exhaust in the air. Jack sighed with relief.

That bikie usually wore white Y-fronts while lounging by the river, stroking his long white beard. He fingered his beard while staring with blue eyes, his Y-fronts twitching. He once had lots of kelpie dogs; they were chained up and the river flooded, and they all drowned.

June heard he liked to cook meth in a caravan in the national park. Growing marijuana was a sideline, and Bam Bam had installed an electric fence. He pumped water from a drying river with a diesel pump, but now it was a dead crop.

That afternoon, after asking around town, June had the story about Bam Bam. She wrote out the details on sticky notes and pinned them up on her office board. The bikie had a relationship with the Hell's Angels chapter at Huntington Beach, California. Bam Bam worked with an American guy renting an apartment in Dubbo and buying meth. June reckoned she could write a true crime book about it.

William told her that people were used to quality drugs, but ice had a devastating effect on society. Sending them all crazy and violent. The gang transported the drugs in the tyres of Harley-Davidsons.

William whispered to June, 'Mays Bend near Bourke was a location for meetings by bikies. They would camp along the river for chapter meetings and chats about white supremacy. They brought their own lounge chairs and drank beer all day. At home, on the internet, they imported automatic weapons like the AR-15, weapon of choice.'

June drove to town and gave this information to Sergeant Blackett. He grunted and placed the notes in a clear plastic bag and tagged it. She had a feeling it would not be of interest, as the copper seemed bored and overworked.

CHAPTER 9

1st March

Aunty June was a brown-skinned figure in a landscape of devastated trees. Her sleep in the darkened bedroom returned her to the old Darling River. Deep green water, luscious and cool, where Yella Belly fish hid in Ngiyampaa and Barkindji country. The Ngurrampaa country called out. She rolled over on her side. The Great Rainbow Serpent Waaway's eye glinted from this river, jet-black sky of stars, where the Milky Way wiggled like a black ribbon overhead in the serpent's path. His pearl tongue flickered around Orion's belt. The serpent's eye looked from black water deep into June's. It whispered, 'I'm Madukka.' In her dream, she cried out: 'How do we live when the plants and animals are extinct?'

June ran her fingers through her gold-tinged hair. She would have to shout at someone to rub out the curse of dreams that coursed like murder though her sleep. Or take a little blue pill. She had seen it all: pillars of water and smoke and burning bushes towering above her house near the Darling River; snakes gone mad; herself running stark naked through grevillea bushes dripping with yellow honey. She shook like she had an alcoholic tremor, but she had never even drunk a glass of cooking sherry. She couldn't get to sleep half the time. Nothing worked, not even late-night sweet masturbation with images of forbidden fruits and that man, Bundi.

'I need a cup of coffee, now!' she hissed through clenched teeth into the kitchen sink.

Uncle Jack was wearing a green two-dollar shop Irish leprechaun hat with a red beard. He was sipping a beer to celebrate St Patrick, he said.

No one responded in the kitchen. They bent their heads as her Aboriginal flag earrings announced her presence. The necklace of echidna quills and emu feathers lay against her neck like a wreath. The rest of her family ate raisin toast and plum jam at the laminate table. No one mentioned the morning madness. Nice manners, please. June made coffee like a nice, quiet, middle-class Murri woman who quite liked eating rocket and kale. The room was small and tidy with white-painted cupboards and a table with a cloth of Tiwi screen-printed kangaroos on linen. A single Ikea light hung above the table.

'Tough night, Junie?' said Merle.

'Usual goona: river full of dead stinking fish, or signs that read "Massacre Downs" or "Gins Fall" or "Treachery Creek",' moaned June.

'We're working on it.'

'Got to have faith: the desert will bloom, and the water run free. And I believe in winning, like at footy,' said William.

'Praise the fucking Lord,' said Jack.

'Language, Unc!'

The whole football-mad family looked forward to the annual knockout competition in Wilga, their outback town. Their true

Country. It thrilled June that her niece Arana would be a debutante at the Black and White Ball after the Knockout.

'We've got six weeks ahead. The Western Rivers community will come together about water rights. Arana better put the info up on her Facebook page to advertise the river meeting.' June leant against the kitchen bench, and she tapped on a laptop computer with two fingers.

William yawned on the lounge with a newspaper stretched on his knees. His curly grey hair stuck up like a brush, and his brown face glowed.

'I made an advertisement for my investigation agency, and I invited everyone to like the river Facebook page,' said June.

'You telling me anyone looks at that stuff? Rubbish,' said William.

'Did you see what someone wrote on the page? "Unless you pull down this Facebook page, Big Cotton will destroy your family and your Indigenous shit. If you don't stop, you'll disappear like the blackfella from the riverbank." What a disgrace, threatening us,' said June.

'Those politicians visiting the river towns had to be escorted by police. It's not just Murries who are wild,' said William, and he coughed. 'We can put up posters, same as Menindee. They never did find young Thommo.' He picked up a stone implement and traced the edges with his thumb. 'Reckon he killed himself.'

'I'm not giving up the search for what happened to him. The gunjie are next on my list of suspects,' said June. 'And for the river's sake. I think someone is trying to manipulate those cops. They seem useless. Every time I go to the Reconciliation meeting, that senior sergeant sits down next to me. I say to him, "Do you have to sit next to me?" He just laughs. But he's wearing his police uniform.'

'Embarrassing, eh?'

'I'm going to talk to them, I can't wait anymore.'

'Mother Earth is telling us something's wrong, that river is all dried up, no kuya, no yingka yabby,' said Uncle Jack. He was hunched on an old lounge chair with a small radio attached to his ear. His football jersey was covered in Aboriginal designs.

William's face crinkled up as he laughed and stroked his salt-and-pepper beard. He had a job with the local council, but mostly he stayed home. Uncle Jack needed care as the Elder; they kept an eye on him, as he sometimes received TV messages from spaceships or Mossad.

* * *

On the mantelpiece sat Aboriginal artefacts, shell art from La Perouse, a Harbour Bridge of periwinkles and blue velvet, and an emu egg carved with Slim Dusty.

The house had a big 'night and day' lounge, which Merle paid off on hire purchase from Harvey Norman. The dining table was nineteen-seventies, with a green laminate top. Merle kept it immaculate except for the piles of posters they made about saving the Darling River.

June's room, a lean-to perched behind the house, had a corrugated iron roof, allowing the rain to tap her to sleep.

'There won't be fish, no Yella Belly or Murray Cod left. The pintuukay mussels and catfish are gone,' said Jack. 'They're stinking dead. Baiame the creator walked from Cobar to Gunderbooka and left his footprint at Byrock. Holy land. All dried up now; bloody cotton.'

'You know, without the river, we're nothing,' June sighed. 'We can try to save this one river system. As a child I swam in the deep water with cormorants and pelicans gliding along. Coolibah gum trees leaning into the water. Like guardians. When they cut down old-growth trees there's no way to soak the land. Creeks dry up.'

'Them gubbahs don't give a stuff. That New South Wales Premier said, "I don't care about fish, just people."'

'She's the Queen's representative.'

'Fuck the royals.' Jack grimaced.

'We are the fish! No water, no fish, no people,' said William. 'We might as well be finished up. The river is our blood, and the animals along the river, they're us, we're the animals. We grew up catching yabbies with a bit of meat on a stick, eh? Ngiyampaa, Murawari and Yuwalaraay people, the heart of that Murray–Darling basin.'

'The fisheries people are trucking hundred-year-old Murray Cod to hatcheries in the south,' said June. 'A million fish dead.'

'Should knock the Murray–Darling Basin Authority on their heads with a nulla nulla,' William groaned. 'It's mismanagement, not bloody drought. Bundi is in the land council, I bet he was doing deals to sell Aboriginal land to cotton farmers and pocket some money. Thommo's sabotage was interrupting sales.'

'Reckon gunjie did it?' said Jack.

'I should study law. It's not enough to have my PI certificate,' said June.

'Why don't you get off your arse and read up on that entrance test for uni, eh?'

The phone rang and June answered. 'Yeah, it's Yanakirri investigation service. Who are you? What do you want? Wait on, tell me about him. You want me to do what? Hello?' June put the phone down and stared at William.

'What's up, sis?'

'He talked about Thommo. Reckons he knows who is responsible. He won't go to police, says they're corrupt.' June shook her head. 'He didn't leave his name or contact.'

''Cause you've blabbed around town for ages asking questions. You'll have to find answers. The gunjies didn't bother – they're in deep with the cotton barons.'

'I took your shovel to the cops after someone left it in the shed,' June mumbled.

'What you do that for? My shovel!'

'It might be evidence.'

'Poop to that.'

'I don't like a mystery much, not when our mob is being set up,' said June. 'I'll keep interviewing everyone who knew about him and his work. You lot might be able to help me, especially that young Steve – he's going to be a lawyer one day. Everyone will be a suspect. Including Duckie. She has a terrible temper when she's drinking.'

* * *

Bundi came up to June and sat next to her on the street outside the courthouse. She inched away, but he grinned and offered her his chip packet.

'What do you know about Thommo?' June said.

'I saw that young fella in a nearby town,' said Bundi. 'He was alive all right, a while back.'

'He used to work for the Department of Roads,' she said, 'but he was fired, then he went downhill.'

'After robbing the Gobblers. A yarndi deal gone wrong.'

'You're making that up!'

'Cross my heart.'

'Why would you do that?'

'I'm just saying I reckon I saw him. I know what he looks like; he's got those big front teeth like Bugs Bunny.'

'He's dead. Police said he has just disappeared or run off after domestic violence. But they are hiding something.' June fiddled with her phone.

'I heard that Duckie was a drug mule at sixteen,' Bundi said. 'Bam Bam gave her lot of expensive outfits and money. Picked her up on a Harley-Davidson. Out at his place there was a long line of utes to pick up dope supplies. Cowboy guarded the crop with a gun and a dog.'

'Yes, I heard,' sighed June.

She felt empty, as though she couldn't believe anyone anymore. But the gossip seemed to go on and on. She felt like a madness was dribbling from the walls, and that she was a tiny figure standing on the four corners of the earth. An existential wasteland of doubt and false leads. No one seemed capable of telling the truth, not even her own people. Not even her own brother. She tried to remember what her cousin looked like. Did he appear on the community website from last year's Knockout? His mother didn't even have a picture taken in the last few years. June knew he had always been in trouble at school, for truanting. He never had a school uniform, and she gave him lunches. Poor dead kid.

CHAPTER 10

2nd March

It was time to interview Sergeant Blackett. June and Merle waited outside the bowling club on Friday night, and sure enough they spotted Sergeant Blackett with his arm around a local Murri woman.

Merle yelled out, 'Hey, Sergeant! We want to have a chat, off the record.'

Blackett looked over and shrugged. 'I'm busy, you girls will have to wait in line.'

But the Murri woman with him scuttled away, laughing. She ran off down the road.

'What the hell do you two want? I'm off duty.'

'Just a few questions about the death of Thommo,' said June as she turned on her most engaging smile. 'I've done the course, I know what to ask.'

'Case closed. It won't ever get solved. Little TAFE course, and you're an expert now?'

'Sorry, Sergeant, to be a bother, but we will ask questions until there are answers. That young man was not mixed up with big drug crime. I can vouch for that. He thought you might be offering to help him get off a charge of interfering with private property after the last "sit in" against gas fracking. Did you meet him that day he was killed?'

'No body, no crime! I didn't talk to him or offer any bloody thing. Your information is incorrect. I don't have to answer any of your dumb questions. I can't talk about police matters. I'm going home.'

* * *

Next morning was quiet. The council workers moved slowly down the wide Wilga street on the back of a truck. They stopped at every flowering jacaranda tree and sawed off the branches. Meanwhile, the kerb and guttering team worked their way along dusty back lanes, concreting dirt while young Murri Aboriginal CDEP workers swept the road.

One man leant on his broom and called out to Constable Johnson: 'Morning, officer. You got a durri for a fella?'

'You want cancer?' said Johnson. 'Keep sweeping, you missed a bit.'

He laughed with them and stopped to chat about football. He was keen to develop his community consultation skills. He was determined to not be a shithead gunjie. He would show the town that he had time for everyone, not just white people.

The main street had a park with some pathetic caged emus and cockatoos. Johnson hung against the wire and sighed. He didn't like caged animals, just humans. A broken swing swayed in the dusty breeze and struggling ghost gums near the main town buildings sang out in shining white. The edifices were historic marvels of sandstone, with government crests of English crowns and kangaroos. The courthouse had been built there by accident; now as a court of law it was the site of main theatrical events, sentencing young Aboriginal men to gaol for offences such as driving without a licence. They were carted by the dozen into court in Sergeant Blackett's wagon, all going away for a holiday of incarceration.

Later in the office of the Wilga police station, Johnson was not feeling well. He might have been hallucinating; maybe those mushrooms on toast were really magic mushies. A stupid joke on him at the Red Rose Café. He sometimes just wanted to run home to his mum in Newcastle. He kept seeing visions of the river all full of bodies, but in reality Wilga was a very quiet town with topiary bushes, hedges and gnomes in people's front yards, an occasional statue of a blackfella standing on one leg with a bent spear. No dead bodies there. His visions of being a top cop were kept quiet.

Inside the office there was an air of hushed expectancy. Sergeant Blackett whispered into the phone: 'Not now, mate. Look, Mr B, if you want to talk … listen to me … usual spot. I care about the irrigators and cotton farmers …' He growled and then laughed. 'Same as you, you black stud, see you round like a rissole and remember, gamble responsibly. Yeah. A big five hundred in a brown paper bag.'

Constable Johnson stood listening by the door, then he walked in carrying the mail.

Johnson was twenty-five and gormless. He had blond hair, freckles, pale eyes and a wide red mouth with white teeth. He looked like a nice lad, with pressed creases in the uniform. He had shined his badge and polished his gun holster. There was a picture of his mum in his wallet. New to the New South Wales Police Force – his first job and bursting with youthful enthusiasm. He missed having a girlfriend, and felt homesick and overawed by his sudden transportation to the bush. He thought about the comfort of his single bed in the men's quarters where he could take out his dick while dreaming of female breasts.

'You catching up with the cotton farming lobby?' said Johnson. 'I thought we had to protect all citizens.'

'That right? I'll flog any law-breaking bastard that touches a pump or gauge. You know what? I don't like eavesdroppers. We're here to keep the law and protect the farmers and the establishment. Property rights. Don't forget who pays your wages. It's taxpayers, not greenies and blackfellas, they don't pay taxes. They're on the dole.'

'Everyone is on some kind of government handout these days, sir,' said Johnson.

'We don't even let the Blacks chip cotton anymore.' Blackett laughed. 'It's cheaper to import Chinese workers. The Blacks don't own the country or the water in the river. No special treatment. We all share it, and we gotta make it economically viable. Sustainable growth for development and prosperity.'

'No, I read about it. The Murri people protected the environment, and they fished for only enough to feed themselves on any day. It was sustainability,' said Johnson.

'What fish? Fish fingers?' said Blackett. He stood up to tower over Johnson, walked away and flicked a pen at the constable, who caught it mid air. 'Well, they don't have a big bloody freezer like I have. It's full of kangaroo tails that I flog to outback stores. That is what I call bloody sustainable! And guess who's buying them?' He guffawed.

'I worry about how low the river is,' said Johnson. 'What about climate change and all that? The planet will heat up by two per cent if we don't stop fossil fuels.'

'Oh, shut up! The low water is cyclical drought. The weather kills us, not the irrigators. It's science! El Niño and all that. Argh, make me a coffee with two sugars, and make it hot this time. Last cup was weak as piss.' Blackett looked down and doodled on the notepad. Drawing a nice white Permalum roof. He really wanted to retire and live in a town house in Woy Woy. Go fishing. Go to the bowling club. Eat a veal parmigiana and chips on Saturday nights. Play the pokies. Play Keno. Watch the footy. Poke his wife with a pillow over her head so he could imagine she was fifteen and dark-skinned with a moist spot. 'You grinding the coffee beans yourself?' he called out.

'It's coming, Sarge. Your wife must get sick of your orders,' said Johnson.

'Yeah, I've thought of transitioning Mavis to retirement by pushing her overboard from the cruise liner. Shining Princess Cruises. Fuck me dead. She booked it a year ago, still paying it off. She reckons she's tired of being married to a stingy arsehole. In her time she was a real goer, experimental, "Knees Up Mother Brown".'

'Enough.' Johnson nodded with a lame laugh.

Blackett dialled the phone again and wrote notes in his booklet. 'Sergeant Blackett here, can I please speak to Detective Brown? . . . Dick, is that you? Sorry, mate, had some pressing local business to attend to. Gotcha. Greenie city stirrers, Environmental Defenders, my pretty arse! Some of those hippy chicks could do with a local insertion. See-through blouses. Bend over, I'll drive. Ha. Locals don't want them . . . Hmmm . . . might sabotage the irrigation dams. What's his name? A . . . what? Spell it . . . All right, got it. Steven Allunga. Cheeky greenie? Asking questions, is he? Hey, you know what you get when you mix a greenie with brown skin? You get shit. No, I wouldn't shoot the lot of them, if I were you. It's against the law. That's it. You too, you bludger.'

Blackett hung up and started some paperwork.

The phone rang and Johnson answered. 'Wilga police station, Constable Johnson speaking . . . You want to what? Oh, really? Dob in a water thief . . . Mmm, what's your name? Merle . . . Yes, I need some particulars. How much you reckon? That much water? Yeah, a broken pump? That's terrible. Cotton Corporation. Ancient trees dead. Yes, thank you, we'll look into it.' And Johnson hung up, looking very interested. 'Shit! What do you reckon, Sarge? That woman says there's a big conspiracy to steal even more water, upriver.'

Blackett turned to Johnson, calm and concerned. 'Johnson, we've had nuisance calls before. There's a bit of a problem in town, a smear campaign against Big Cotton up the river. I checked it out – it's nothing to worry about. The water gauges are foolproof and it's all legal. That Merle woman's a stubby short of a six-pack. You've got a bit to learn, leave it to me.'

'Whatever you reckon. Hey, Sarge, what's this about?' He pointed at a newspaper article on the noticeboard.

'It's nothing.'

'Ronald Bruce Blackett, writing for the *Wilga Gazette*? A poem! Is this it? Can I have a look?'

He read aloud as Blackett recited the poem silently to himself:

> Ode to the Nyngan Flood.
> River snaking through red dust,
> in the good earth, the man all trust.
> Through rain and drought, the golden grain,
> is life and breath and worth the pain.
> The Nyngan folk fight tooth and nail,
> against the wet and blinding hail.
> Sometimes, in their darkest hour,
> they fear deep rivers and all their power.

'Ah, it's nothing,' said Blackett.

'And there's a piece about missing explosives from the Department of Roads. Over time, they lost heaps. Are we investigating? It says possible terrorist links.' Johnson began to sort the mail. 'It says June Thomas is suspected of having information.'

'I'm looking into it. Probably those city enviro protesters, we should run them out of town.'

'Lot of Blacks downtown, coming by the busload.'

'All boozed up? Poor buggers can't handle it.'

'Nope, just getting their families off the bus,' Johnson said. 'They're good footballers.'

'Yeah, good for something. You know, it's not true their brains are smaller than ours.'

'Yeah, good at hunting too.'

'Reckon you know all about them, do you?' Blackett retorted.

'Well, I'm from Newcastle, played footy with them.'

'Think they'll respect you because of that?'

'I don't know. I know some families and I've learnt a bit about their culture. I did Aboriginal Studies at the Goulburn Police Academy. Really interesting.'

'Crap! You can't tell me anything about Abos I don't already know. The last young, caring and sharing constable got glassed – twenty stitches, trying to stop some blackfella beating up his missus.'

'Geez, Sarge, not all of them are like that. I know some terrific Aboriginal people up home,' said Johnson. 'That Aboriginal Liaison woman's got some good ideas on Black–white relations.'

'She knows nothing. It's all out of a bloody textbook. I know these people – they're like kids, and you gotta take care of them.'

'I don't know about that,' said Johnson nervously.

Blackett lost his temper. 'I try my bloody hardest at the Police Boys Club, running boxing for the Black kids. They got it all over the white kids, but they end up hitting the grog and it's all over – and the government still gives them bloody handouts. Black rights? How about the poor whites in town? Where are their special privileges?'

The fan turned slowly in the languid heat and flies gathered on the limp paperwork.

The office phone rang and Johnson answered. 'Mrs Thomas, you say there are suspicious blokes sitting in cars outside your home? We'll look into it.'

'She's a troublemaker,' said Blackett.

'Might be, but she seems pretty smart.'

'They always send me the ones with fatal flaws. Come on, son, tell Uncle Bruce the truth. You're a fruit, aren't you?'

'I had a girlfriend in Newcastle a while ago, not that it's any of your business, sir. Times are changing you know – it's not a crime; some of my friends are gay,' said Johnson.

'Keep your bra on, I'm only kidding you. Newcastle, eh? Too far to go for a root, mate. You need a local squeeze. What about

Fay from the hairdressers? She's Italian, nice backside, single, attractive. Or your old girlfriend, Duckie?'

'No, she wasn't, it was one date. Anyway, Fay's thirty, she's got a moustache, and I'm twenty-five.'

'She might teach you a few things, she would be grateful, and she might have a tight one,' says Blackett.

'I'm not like that.'

'We're all like that. No point wearing out your fist with wanks.'

After that, silence. Johnson and Blackett continued to work on endless reports, the atmosphere was smelling of something dead, as if a rat had died behind the gun cabinet.

CHAPTER 11

4th March

The household woke early and everyone fell for William's joke that a truck full of lollies had tipped over down the street. He handed them saucepans to run out and collect the loot.

Arana looked disgusted. 'Dad, you do the same joke every year.'

June went down to the Reserve in the late afternoon to visit old Mary in the tin house by the Darling River. On other days she had watched Mary fill a pipe and rub the baccy, biyagaa, in old, gnarled hands. Her face was a hundred years old; her parents had been traditional people. Her initiated father knew the old ways but didn't teach Mary her language. She remembered a smattering of words. She liked the young ones to catch her a goanna, mangun-gaali. The old lady didn't like anyone to drink grog nearby. No swearing either, so Jack had to look out or she would hit him with a walking stick.

June sat still. She almost held her breath sitting on a wooden box outside Mary's place, the corrugated iron walls rusting. She thought how it must be stinking hot in the day, but tidy as any fine house. Old Mary was not home; the place was deserted.

Later, June sat by her car in a shady parking place used by night-time lovers. It was pretty, with ancient trees. A fancy car drove up and handsome Bundi stepped out.

He saw her and shrugged, placing his hand on his heart. He winked and sighed. 'Did the police talk to you about the explosives?'

'I said I knew nothing.' June hung her head.

'Sure, nothing, but there's rumours. Duckie reckons you might be involved in a terrorist plot.'

'She's a bit of a fruit cake.'

'I told Blackett that.'

She watched him undo a cigarette packet. He smoked but put it out after two puffs. Her chest felt tight. A bit anxious. She wanted to talk but was unsure if she could trust him. What if they found evidence about her on Thommo's phone? She regretted that it was given to the gunjie. No one knew about her plans for activism except Duckie.

June gazed at Bundi's throat, shining like liquid gold. Her fingers twitched with a need to stroke his cheek and lips. As he smiled, the dark skin revealed red inside his lips, and she sighed and shuddered. He was very beautiful, like silver wattle along the riverbank or a strong red river gum tree. His broad chest pushed against the blue cowboy shirt; the buttons were mother of pearl. His dark eyes glinted, and his crinkled face lit with a broad grin. June ached to lick that pink inside flesh, just to taste his mouth, just once. She wanted to utter something like, 'Look at me, I'm waiting for you, kiss me because I can't stand it anymore.' She noticed she was wearing her worst t-shirt, with a hole on the breast, showing her flimsy bra. Her nipples stood up.

'I should go,' she said.

He moved forward and held open the car door for her. She had to pretend that it was all normal and her burning face was just the hot flushes, but his cologne was mingling with fresh male sweat. Since when did Murri men spray cologne on themselves? She pushed her nyamus up and he noticed and nodded as if to say, 'I'd like a piece of those in my mouth.'

Then she stopped, turned and put her hand on the fly of his jeans. She gripped his belt and gave a little tug. Her hand next to his nhiimay, and his eyes met hers under the black lashes. But he mentioned something about rain or dust or fire or getting a live sheep for a barbeque. June couldn't stop looking into his eyes. She felt electrified, and looked at his brown leather belt and silver buckle, embossed like a rodeo prize. The hot naked skin of his arms where the sleeves rolled up. She gazed at the signet ring of onyx, and his large, manicured hand.

His mouth twitched and didn't speak. They were frozen and staring at each other. He breathed out and she ran her fingers down his groin. He clenched his lips but acted like it hadn't happened, that it was getting late. He looked down the riverbank. She let out a breath and he pinched her chin with his thumb and fingers and nodded and stepped into his car and drove off with a wave. She had to sit down, right on the stone ground, with her hands pressed between her thighs. She thought, I'm so bad.

June returned to the Reserve and unpacked food for old Mary, who soon walked into her house. June made tea on the fire inside and took down the precious tea set to pour the drink. Willow Fashion plate. A spoonful of powdered milk and some sugar, and June set the cup aside to wait for the old lady to talk. June wanted to ask her about the old days, and she was ready with a notebook in case Mary needed her to write down her stories. June listened to the night birds.

'Bundi, eh?' Old Mary giggled.

'Nothing.'

Mary beckoned and sat in her chair, listening to crickets. She lit a kerosene lamp as June set up the sound recorder. 'He hot all right, that one, but big trouble … You want that story, old one from my birthplace? That Gwydir River place?'

June nodded. 'Yes, Aunt.'

Old Mary looked like she remembered standing by that sparkling river. June waited with drawn breath and then Mary hung her head, and a big tear ran down her old face. June had a sudden thought not to ask this question. So many had already asked. She held out her arms and listened to both their heartbeats.

Mary told June her story, smoking her pipe and beginning. 'That Gwydir creek, it be some terrible, painful truth,' whispered old Mary. 'In true one place, a big mob, maybe fifty people, they were shot down and burnt.'

'Why hadn't anyone told us all about our history at school? Were you supposed to keep it a secret?' asked June.

June's family living not far from a spot where terrible murders had taken place, the soil full of human fragments, skin and hair and terror and murder – bloody, horrifying murder. It was a kind of stinking trickle, a miasma oozing from the grey earth. You couldn't see it or breathe it in, but it would never go away – it was like a river, like blood coming from below, the pitiful dead risen from the ground, burnt skeletons dancing on graves, like a devil-devil dance, crisscrossing red footprints across the escarpment, searching for their loved ones, the spirits searching for home, all the time crying. Their heads were placed on spikes and bodies hung in trees; entrails hung, with carrion eaters tearing them to pieces. No wonder the winds howled. How could anyone stand it? How could they go on living? And the people in the local churches still sang 'save the Lord'.

The old lady tapped her pipe and refilled it.

'It was the Gwydir River in New England where that killing took place,' said old Mary. 'It was our Gamilaraay people killed in 1838. That Waterloo Creek, it like a bloody scar, never go away, you know? Settlers attacked by our fella great, great warrior grandfathers. Our land got stolen. So the Governor of New South Wales sent off that Major James Nunn. There's a monument to him somewhere, a hero, so all the whites reckon, but he a murderer all right.'

'We could tear down that statue. I suppose they wanted to teach the Blacks a lesson,' said June.

'Too right. That Nunn, a frontier soldier man who led a group of more than twenty troopers to that Liverpool Plains, and he make sure he kill every Black he met. They got horses – yarraman we call them. They want pumali kill them. And the men's chests, they crisscrossed with bullets like they were heading off to face war. Them saddlebags carry lotta strings of bina dried human ears, so he can count them up, see what number of dead Gamilaraay peoples. No witnesses, no one to tell. That statue is in one of the towns, next to the RSL. You can pull it down, eh?'

June leant forward and adjusted the recorder, and old Mary smiled at her and nodded.

Then June made a cup of tea for old Mary and tidied up her recording equipment. 'Some bird gives signs, galaathil' said old Mary. 'Crow, old waagaan up there, means it rain baarri-y real heavy soon. You know I want to make sure people are punished if they break our laws, like men who destroy our burial sites. That Thommo bulldozed one.' She nodded at June and it started to sprinkle from a cloudless sky. A willy wagtail, thirrithiri a messenger, you didn't want frolicking in front of you. Death was coming to someone. Old Mary sat on a new red blanket and finger-knitted dillybags. She held June's hand and stroked it, and the old woman could read her spirit. June kissed her forehead and drove home.

CHAPTER 12

6th March

The Wilga Railway Hotel was an old pub with wide verandahs, wrought-iron balconies and pillars of wood across the front. The local whitefellas strolled around town and went to the supermarket and the stock and station agency and the RSL club next to the war memorial. It had a metal-cast soldier with a rifle, and a plaque with decaying flowers on the granite base. There was no memorial for the thousands of dead Aboriginal people in the Black Wars. No recognition that Major Nunn had murdered hundreds of Aboriginal people and left them in burnt piles for crows to eat. No Waterloo Creek plaque. No Massacre Falls plinth or Gins Leap statue. Just smiling bronze, white men on horseback, ammunition strung across chests.

The whitefellas hung out with their wives at the Wilga Bowling Club and ordered the T-bone steak with chips and a beer while the missus had a chicken schnitty and a white wine. No Blackfella would usually go to the bowling club unless he was an out-of-towner Koori from Sydney, trying to prove a point and flash his or her credit card and order the prawn risotto.

The drinking troughs for horses were out front of the hotel, with the occasional horse tied to the railing. Every night the place was full of men drinking beer. It was a real outback hotel, and a real outback place of segregation and blatant racism. The Aussie way. There was no particular sign to indicate this apartheid, just some jokes in the form of cartoons, all dog-eared and stuck to the wall above the bars. 'Gin piss' with a picture of a black woman naked. Nice stuff, really funny, if you were a whitefella.

At six o'clock, it was time for Happy Hour at the pub.

'Come on, I'm tonguing for a charge!' said Jack. 'It'll take the cough off my chest.'

'One beer is all we're having,' said William.

'You want to come, Rara?' said Jack.

'Nope, I'm studying, and my boyfriend Steve arrives soon. I got to wash my hair.'

There were two unofficial bars: one Black, the other white. Blackett and Wayne Johnson were off duty and in the white bar, drinking, while Jack, William and June were in the Black bar, but June had to go over to the white bar just to prove a point. Arthur the publican stared at her, as usual.

'Just ignore me,' June said.

'Unfortunately, June, you are not an easy person to overlook. You are a handsome Murri woman, and you stand out like dog's balls in my nice exclusive gentleman's bar, where if you can read the unwritten signs, women are not allowed, no matter what colour,' said Arthur. 'You are supposed to drink a light beer, chardonnay or a shandy in the ladies lounge.'

'So old-fashioned! Well, blow me down, I never knew. You see I was expelled from primary school for being an Aboriginal person, so I didn't learn to read with proficiency. Just give me my drink of lemonade and I will sit quietly.'

'Look, I have tried being polite. You're a respectable lovely woman who owns a house uptown, but rules are rules. Out you go, June, or I will have to call the constabulary,' said Arthur.

'Well, you usually set up that gin jockey sergeant with one of our young girls, so why don't you call him over to remove me?'

'Just go, will you? Give me a break, I'm trying to make a buck. Move,' said Arthur.

'I have rights.'

'You don't even look Aboriginal.'

'And you don't even look racist.'

He wiped over the bar and squeezed the cloth out. He arranged the beer mats and stubby holders with bare-breasted girls on them in a neat row.

June drank slowly and smiled her toothiest smile at him and waved a straw. 'You know, banning plastic straws and glasses would help the environment. Do you know anything about Thommo?'

'Nothing, except he was wanted by some heavy blokes for some deal gone wrong. Try asking around the shire council. But don't tell anyone I told you! Rumours get you killed around here. You better be very careful. I hear everything in this bar. He was mixed up in environmental actions and dealing marijuana, but small-time.'

'Not enough to get himself killed. Gunjie don't care.' June sighed.

'They're in it up to their necks, make money on the side.'

'Maybe Big Cotton put them up to it?' June whispered.

Arthur shook his head. His eyes flashed as he noticed Sergeant Blackett tapping his nose at him.

Arthur went very quiet. 'Don't start spreading that around town. Be careful!' he whispered. 'I mean it! You might be next. Those corporations are now worth countless millions. No sniffy little Murri fella would have threatened them.'

June took hold of the stubby holder with a bare-breasted black woman on it. 'I reckon I might throw this one out – it's sexist and racist. I find it very offensive.'

The barman grabbed June's arm.

'Arthur, let go. You're hurting me. Who were the heavies? What names?' said June.

'Don't know my own strength.'

She dropped the stubby holder. He wiped the beer mat and nodded at her.

'Your beautiful dark bosoms, your nyamus are what I dream about,' whispered Arthur.

'Praise the Lord! It's sexual harassment. You could also remove that Eureka flag from the wall – it's come to symbolise patriotic right-wing rubbish,' said June.

'Whatever you say.'

'I might go and have a drink with a better class of people over there,' said June.

'That would please me greatly. Anytime you want to go out for a meal, just call, okay?'

'Get your hand off it, Arthur,' laughed June, and she walked in a very dignified manner to join her family in the other bar.

William and Jack watched June stroll back to them and sit down on the bar stool.

They slapped her on the back. 'Go sis.'

'Another drink?' said June.

'You shouting?'

June reached for her purse, but thought better of it and let William pay.

'Got a spider in that wallet?' he asked.

'You seem to have a lot of fifties in yours. Hope you're keeping the river fund money separate?'

'What do you reckon?'

'Just asking,' said June.

'I gotta go down the Aboriginal bank,' said Jack. 'The TAB, eh?'

'Lotta deposits and no withdrawals. See a lotta Murries walking round Wilga with stiff necks from watching that TAB screen,' said William.

'Who's gunna win the Knockout?' said Jack.

'It's a mug's game, betting.'

'Up the Warriors!'

'They should have put that player outside. He didn't know the rules.'

'Did you see him score?' said William.

***

Over at the white bar, the off-duty police were about to order beers as they leant on the counter.

'I'll buy the first round,' said Johnson.

'That'd be a good start to your career, son.'

Arthur poured their beers.

'What's it like working here, Sarge? Seems a bit quiet.'

'Argh, every second Thursday she gets a bit rough – pension day.'

'Yeah, why's that?'

'Government handouts. They all get down the pub, and I gotta come down here and straighten them out,' said Blackett.

'I still reckon it must be a bit of a bludge: so quiet.'

'Look, I've been here over twenty years and I've seen more blood than you've had hot dinners. Take that big bus crash, bits of bodies all over the highway.'

'Sorry, Sarge, I meant it looked like there wasn't much to do at night. The girls in town act like I got bad breath. Hey, another two schooners, Artie!'

He held up a hundred dollar note, and Arthur shook his head.

He put it back in his wallet. 'On the house, is it?' He smiled and leant over the bar towards Arthur. 'You haven't got an alpaca out the back, a sexy bit, have you?'

'The one that escaped from the Gobbler motorcycle gang clubhouse after a few had a go?' Arthur laughed.

'What was she wearing?'

'Inappropriate attire, in wool,' said Blackett.

The men guffawed.

'I wanted to date a local,' sighed Johnson.

'Yeah. Well, your profession goes against you in the social area, but there's the bowling club, got a disco on Friday nights.'

'You'd be a bit old for that, wouldn't you?'

'Ah, I go round and see if anyone needs a lift home, you know?'

'Aw, yeah,' said Johnson.

'Nah, I keep pretty busy. I play the Hammond organ for the church fellowship, you know, keeps Mavis happy.'

'You still have relations with your wife?' said Johnson.

'Cheeky blighter. I'm right in that department. I sometimes wish my libido would just relax. It can't be normal wanting to back-door young girls,' laughed Blackett.

'Too much info, Sarge,' said Johnson. 'You know, I was out bush and stayed at a pub where the owner had a reputation for liking horses and dogs. He was walking up the stairs behind me and he said, "You got a nice arse – if you were a kelpie, I'd root you." I got out of there quick smart,' laughed Johnson.

'Well, you would,' said Blackett.

Johnson looked over to the Black bar. 'What goes on over in the other bar?' he said.

'Maybe you should have a look, eh, son?'

'Might do that,' said Johnson. He walked into the area, and he recognised William and nodded and grinned. Jack slouched low in his seat.

'All right, I'll have a Foster's. Don't tell Merle,' said William.

'One won't kill you, brud,' said Jack.

The barman put the drinks down in front of the men and William downed his in one gulp.

'Jeez, thirsty or what?' said Jack.

Next thing, Johnson had pushed through the other staring drinkers. He stopped in front of William, eager as a puppy. 'Hey, William! How ya going?'

He held out his hand to William, who failed to recognise him for a few seconds, then June nudged him.

'Where do I know you from?' William looked at Johnson with eyes focusing without his glasses.

Johnson leant in. 'Newcastle. You know me mum, I'm Wayne.'

A grin lit up William's face. 'Newcastle kid, all grown up and, for God's sake, a gunjie, a policeman. You're Wilma's boy. Jeez, you've grown up, eh? I used to play footy with your old man, Joe. How you going?' said William.

They shook hands and nodded as Johnson grinned and shook Jack's hand as well. Jack wiped it on his shirt. The room seemed full of thick expectation. Jack slid off his seat but June pushed him back with a soft hush.

'I'm good, I like being in the police force.'

'Well, we'll be seeing a bit of you, then.' June laughed.

'I don't remember Wilma,' said Jack.

'Nah, you were in the boys homes.'

Blackett stared from the other bar, looking disgusted. A few men in blue lapels crowded around his back to have a look.

Johnson sat on a stool next to his new mates and gestured to the barman that he would buy William a drink.

'What are you drinking, William?' said Johnson.

'Fanta.'

Jack coughed and moved towards Johnson. 'You hear about when I was working as a jackaroo in the Northern Territory?' he said. 'I got one for you. The boss says, 'Jackie, you can take the truck out on the Arnhem highway. Lotta feral animal there. Buffalo and that. But if you hit a pig, stop the truck and hack it out of the bullbar with this axe." I says, "All right boss, but what do I do with his helmet and motorcycle?"'

'Oh, yeah,' said Johnson as he stared into his beer froth.

'Still do a bit a hunting?' said William.

Johnson drank down his beer and wiped his mouth. 'Yeah, I brought my pig dog, Alfie, up with me.'

'Jack and me, we're going out next week early, for a roo. Leaving before light. We could use a dog. Could you bring him round?'

'Righto.' Johnson swung around to face June. 'Who's your mob, Aunty?' he said. 'I know a lot of Wiradjuri people.'

'They're everywhere, all over Sydney, breeding like rabbits. It's very fashionable,' said June.

'Have the police found any more info on the missing man, Duckie's fella?'

'They reckon he went protesting up near Queensland. You could do us a favour and look at the report at the station,' June said. 'We asked but got nothing. You see, Thommo is loved – he's like a son to me. Always helped Aunty Mary when she was in trouble. He loved his little boy, young Billy – he is still waiting for his dad to come home with the toy he promised.'

'Yeah, his mum Duckie's a real beauty,' said Johnson.

'She's heartbroken at the disappearance. Our community don't like a mystery like that, we don't forget people ever. It's very important that he's found. The Murries won't stop looking for him. He made a difference to everyone's feelings about the water being stolen – he brought a political commitment that we didn't have before. People are scared for him. Look it up?'

'Against regulations. Sorry.' Johnson froze and played with a napkin.

'We got Ngiyampaa and Gamilaraay mob, eh, June? Dad and Mum were river people,' said William.

'That's nice,' said Johnson.

'It's one tiny report – the people need to know what happened.'

Johnson hunched his shoulders.

'How long you here for?' said June.

'I'm stationed here for a year. I heard you have a detective agency. That's brave, eh? I'm just having a drink with me boss, the Sarge. He's a good bloke, I'll get him to come over.'

The group laughed nervously, and everyone's head was low to cover some awful embarrassment. Wayne Johnson stood to move over to the other bar, but June gently pulled on his arm.

'No, I wouldn't worry about that,' laughed June.

The sounds of poker machines jingled in the background with men's laughter. For a moment nobody spoke, and Johnson looked out of place and awkward. He fiddled with a stubby holder with a picture of a riverboat steamer and nodded. 'No water for those boats now. It's a pity, the drought and climate change stuff.'

'Water thieves took it,' said June. 'The riverboat company went broke. No more tourists.'

'Stealing water, like a *Mad Max* movie, ha?' said Johnson.

'It's real. Have you seen the muddy puddle at Wilcannia that passes for a river?'

William pushed his stool back from the bar to take a good look at Johnson. 'The tourists had to get out and push the boat off the mud,' he said. 'It would have been funny with them all covered in grey slush, except it's a tragedy that the river has gone, and my daughter is fighting to make us all wake up.'

'It's a drought, the rain will come and flood you all out next,' said Johnson. 'And us police will be sandbagging alongside you.'

'Not drought – capitalist greed of irrigators. There are some of them standing in the other bar next to your sergeant,' said June.

'All cosy in bed together,' said William.

'Rubbing each other's widgets,' said June.

'Old Blackett bending over …'

'To count the cash,' said June.

'Blood money,' said William.

Jack looked up to see Blackett staring at him and pointing his finger at something.

'I better get going now,' muttered Jack as he shuffled out to the toilets.

'Hey, Johnson!' Blackett called.

Johnson looked over at the sergeant, who stood with his hands on his hips and a look of consternation. He beckoned.

'Aw, I'd better get back. See you when I bring my dog around,' said Johnson, and he looked worried.

Johnson rejoined Blackett at the white side and drank his flat beer. The two of them stared at their drinks.

'Don't you ever go drinking in the Black bar, son. They don't want ya. Gotta respect their place.'

'Sarge?'

'That June is sniffing around asking too many questions. She'll get in trouble all right if she doesn't shut up.'

Blackett had a red face, and perspiration dribbled as he wiped his forehead. The room had gone very quiet, and some men looked with vague amusement at the police. The bar seemed stopped in time; then the drinks began to flow again and the noise increased.

A well-dressed Bundi walked into the cow cockies bar and ordered a beer. He nodded at Blackett and raised his glass. Behind him stood Duckie in a black t-shirt and tight jeans. Her eyes burnt into the sergeant's face. He laughed. Bundi fitted in anywhere; he had style and charisma. And the young woman with him was lovely.

Blackett used a soft but fierce voice: 'Look, Johnson, you don't want trouble if you can avoid it. Stay away from that one.' He indicated Duckie and Bundi. 'I have a good life here. One wife, one house, one going-out suit. We don't need trouble or too much stuff to look after. Life should be simple. The Blacks have all they need. They even have their own bar to get pissed in if they want. Our society doesn't discriminate. The land rights stuff, well, what would they do with all that land?'

'It all belonged to them for seventy thousand years.'

The men in blue pure-cotton lapelled shirts at the sports end of the bar burst out laughing, and a look of embarrassment passed over Blackett.

'Give us a break,' Blackett said. 'They would get land rights and water rights, then sell it all. Then piss the money up against the wall ... Goulburn Police Academy, for God's sake! What did you learn? How they lie down all day, get the dole? How to get pissed and punch your missus's lights out? Aboriginal Studies. Stick to your own kind, son.'

Suddenly Duckie was beside Johnson, and behind stood Bundi. She grabbed Blackett by the shirt; she was a bit drunk. 'You're not doing your bloody job! You didn't investigate my husband going missing. You know something and you are lying like all of them!

You're all in bed together: cops, bikies, bloody irrigators and the council.'

June moved over and held on to Duckie before she got arrested. 'Okay, sis, calm down, we're all good here.'

'I look after this town, I have respect,' said Blackett. 'I don't need a drunk sheila attacking me in a public place. Let go or I'll have you up for assault!'

June pulled Duckie away and pushed her out the door.

The barman handed another two beers to Blackett and wiped the whole bar down with his Wettex. Johnson stared into his beer.

'Where were we? Yeah. Come around to our house and Mavis will make you some nice pancakes,' said Blackett.

'I don't know why she said that, sir. I might have said something to her.'

The two policemen walked out of the bar, and Johnson nodded at William. June returned to the bar and noticed Bundi, who kept drinking and laughing, surrounded by farmers.

June tapped Bundi on the shoulder. 'Come outside, I want to ask you something.'

He grinned and winked at his mates as he followed June to the beer garden.

'You were the last person besides Duckie to see young Thommo,' June said. 'What happened that day?' She tilted her head and fixed her eyes on his.

'Nope, I never even saw him.'

'Was he dealing serious drugs? Are you mixed up with Wright? Tell me the truth.'

Bundi breathed out a long sigh and nodded his head. 'I saw him on that morning, but it was a coincidence. Them both heading out along the main road. I was bird-watching for endangered parrots. I had my binoculars.'

'Pull the other one.'

He leant towards her and removed a hair from her face, and she flinched.

'I wouldn't hurt anyone, honest!' he said.

'The family need to know what happened. I saw his tyirr spirit, I swear! But now where is he? He can't have just evaporated.'

'Men run off from violent women every day. They get sick of getting a flogging and everyone siding with the females.'

'Rubbish, you know all about it. You're too much of a coward to tell us.'

'Junie, darling.'

She turned and walked out of the pub. She was fed up with all the liars and sneaks in the town. She watched as the sergeant pulled Johnson along beside him and pushed him towards the police vehicle. Young men were doing burnouts and doughnuts in their cars down the main street, but the coppers ignored them.

* * *

Later that day, back in the budget motel that doubled as policemen's quarters, Johnson lay on his bed and shuffled a pack of cards; then his mobile rang.

'Blackett here, you want to pick up some extra dough?'

'I don't understand,' said Johnson.

'You listening? Money for jam.'

'I don't know, Sarge.'

'Grow up, Wayne. Just agree,' said Blackett.

'I don't think—'

'Eff you, I'll take the money. God knows I need it for the cruise.'

'Sorry, but—'

'Don't snivel. I'll talk to the consortium, find another bloke to guard them. I really don't know what I'm going to do with you. Useless.'

'I won't break the law.'

'Really? We'll see about that. You're wet behind the ears. Let's see how you go under pressure.'

The phone went dead.

CHAPTER 13

8th March

Two days later, June couldn't sleep. She was too hot or too cold or too sweaty. She had cried herself to sleep. Then it appeared – a white smoke in her room. It wove around her and she saw the face of Thommo in the misty shadow. Her skin was cold and clammy, and her eyes blazed with fear.

She whispered, 'What you want, cuz? Let me help.'

The sight of him was enough for her to crawl into a ball under the sheets. She could sense a thudding on the wooden floor and smell his hair. Feel his hot tobacco breath. He pressed against her back and she shivered with chattering teeth. A moan came from her chest and tears welled up. She begged him to leave her alone. She would try to find out about what happened. He was dead and she was responsible for finding the truth for his little boy. June lifted the covers and the white shimmer had not gone. Would she have to live with a goonge in her room?

She leapt up and flicked on a torch. 'Please go away, I will do something.'

A while later, she heard a noise out in her yard – someone or something moved about and bumped into a sheet of tin. She breathed deeply to calm her heart, then quietly walked through the sleeping house and looked into the gloom. A torchlight flickered around the garden shed. What was it? Some kid stealing potting mix. A voice inside told her to stay still, very still.

She hid as a black shape in a hoodie jumped over the fence and ran off. Her breathing sped up and she was in the shed with a match alight. Everything looked the same, then she saw it. The shovel leaning against the bench. William had complained that some bastard had taken it, and even if he originally nicked it from the road gang at work it was his property now. She touched it and noted the clean washed wood of the long handle. Had the cops returned it? It was so strange. It smelt of eucalyptus oil.

June hid the shovel with its incriminating Department of Roads insignia.

* * *

June spent hours at the Wilga police station, but the inquiries about Thommo's fate went around and around. The police kept saying that they had no evidence except her statement, and she was not a properly qualified investigator. There was no proof he was even dead. Sergeant Blackett said about 12 per cent of homicides remain unsolved. He gave her papers about 'efficacy of the police in identifying and charging an offender'. DNA and homicide squads were not warranted until a body was found. He reckoned he would not be giving time to a case that seemed more like a potentially missing person. He would pass on information when necessary to the National Homicide Monitoring Program.

Sergeant Blackett cleaned his nails and shrugged. While the sound of a mysterious homicide was fascinating to her, it was just her word, anecdotal information with not a skerrick of evidence to be found. He said there was an increase in stranger homicides that were difficult to solve. In the end he escorted her out of the station and suggested that the lad had run away and she should not be concerned.

June wanted to yell at him, 'I know he is dead; his goonge spirit came to me!'

The sergeant closed the chain-wire gate and waved her off.

***

June drove out to the scene again. Merle was with her because no Murri woman goes anywhere alone if she can help it. It was hot and very eerie. Merle would not get out of the car; she sat reading with the air conditioner going.

June walked onto the pathway. She swept her hair from her face and her hands shook. Her mouth bone-dry. She kept turning around if she heard anything. A crackle of leaves or breaking stick. She wished she had a sidekick like Vera on TV. The suspense was killing her. In the distance she heard a motorbike. She froze. The roar faded, and for a moment she thought of Merle alone in the car. She breathed in and out and steadied herself against a tree asking for it to guide her.

'I want to come into this Country, wandaba spirits please welcome me!' she called out.

There were days when she felt the crack of a shovel on her skull. Her eyes darting upwards in a dream to see the glinting metal crashing down; her hand flew up to protect, but too late. To punch out or squeeze a fist or run. Like a camera in her head, it was a blinding red flood inside her eyes.

Black crows pecked along the track and eyed her with yellow specs. She had found out about blood and how it coagulates, so no evidence had been found after the bulldozer had cleared the scene.

She climbed into rock crevices overlooked by police and searched for spatter marks. Spindly gum trees with scribble hieroglyphics hung nearby. The track was sand, ochre rocks and dried mud. She looked for anything that might be evidence. Up high on a rock ledge, the kind where the old people placed human bones, she saw something covered in flies. It looked like bone fragment, as if small bits got stuck on the metal and flew off as the weapon was wielded repeatedly. Ugh. She used tweezers to pick it off the sandstone. Then she spotted spatters of what could be blood in dark red on the stone. She photographed it on her phone and sent it to Blackett. Then she placed the dark flesh bit in a plastic bag. Human tissue, surely. It might be a small fragment of nasal bone. She shuddered with the horror of it, and the goonge surely coming soon to get her. Flies swarmed around her face. She drove back to town.

June placed the plastic bag with blackened flesh and bone on the sergeant's desk.

'Yeah. It's evidence you reckon?' said Blackett.

'Genetic analysis can separate animal or insect contamination. This could be a small piece of human bone with protected core marrow. It needs DNA analysis. But if it is Thommo, we are the custodians of our human remains.'

'Looks like a bit of leather.' Blackett grabbed the bag and placed it in his drawer and turned the key. She imagined it sitting there till the end of the world. 'That all?'

June's eyes did not leave his face. He looked sick, sweaty and tired. 'I'm not telling you how to do your job.'

'Looks like it to me. Okay, if that's all, Miss Investigator.'

June heard nothing again about the remains that she was sure were crucial evidence.

* * *

Back home it was time to welcome Arana's young man, Steve.

'Where you go today?' said Jack.

Arana carried a rolled-up Aboriginal flag and placed it on the mantelpiece. 'I marched in the water and culture rights rally up in Bourke.'

'What rally? You should be home,' said William.

'You weren't,' said June. She could see he pictured his daughter in a black t-shirt with a radical bearded greenie nearby trying to get a look at her backside.

'Big mob there, all the Murri and Koori organisations,' Arana said.

Jack smiled and waved at Arana. 'You told them a thing or two? Jarred them?'

'Thought you were coming,' said Arana.

'Nah, too busy. Ran into them fellas from Moree.'

'Had a drink, true, eh?' Arana looked disgusted.

'Give us a break . . .' Jack said and leant back on the couch. June saw his betting slips on the floor.

'You been betting?' she asked.

'Had to pay for the ball dress,' said Jack.

There was a lovely moment when the dress had appeared on the rack in the Red Cross dress shop. Merle and June had sighed in unison to see a white satin sweep of skirt. This dress was a symbol of the changing lives of their young ones. A debutante, for heaven's sake. June didn't say a negative word. It was Arana's life, not hers. All the kids at school were debs.

'How was the meeting?' said Jack as he made his tenth cup of milky coffee.

June looked at his t-shirt riding up and his belt with a rodeo rider. He dropped four spoons of sugar into his cup.

'Some deadly speakers, mostly women,' Arana said. 'Steve gave a great speech about town council and police corruption.'

'Taunting gunjies now, is it? Call that smart? Reckon I got bail money? That Steve is a worry.' William was alert for signs that his little girl might be in love.

'She's grown-up now,' June said.

'I've been seeing Steve. I think he's beautiful, he's got soft hair, thick eyelashes. He knows about Black issues. He's coming to the deb ball. Playing in the Knockout, too.'

'He's leading you astray. Jack, give us a light. What kind of tobacco you got?'

'Gone, I wish it was White Ox.' Jack tipped the pack up. 'Bra, lend me ten.'

'I've only got five,' said William.

'Well, give me that and you can owe me five.'

'What did you do with the ten I gave you yesterday?'

Jack whispered, 'Bought a drink. I was tonguing for a charge.'

'What did you buy that for? I'm giving it away,' said William. As a kid, he waited outside the pub for some white or Black father figure to get boozed up and belligerent. Big sis June grasping the can of Coke and a packet of chips, foster brothers eating most of it and nothing for dinner. They licked the packet and sucked the lemonade straws.

The front door banged; Merle arrived home from teaching at the primary school. June took her heavy books off her and watched her fling herself down on a chair.

'Duckie is coming here to talk to you, June,' Merle said. 'Wants to know if you've been asking around about her missing fella. She reckons you can help her.'

'He's got the same white grandfather as me. He really cared about the river,' said June. 'I can't sleep for worrying about what happened to him. He talked to me about some deal; he was certain that we could force the cotton corporation to return the cultural flow.'

Everyone went quiet, not wanting to talk about that young man. June nudged Jack to lift his feet and sprayed Spray n' Wipe.

'Steve's coming around tomorrow.'

June placed the notebooks on the mantelpiece, but William was alert. 'I don't want her involved with Black Power types or missing persons, bad trouble,' he said.

'He's okay. No one had a bad word,' June said.

'One moment they're telling us we need the bloody Murray–Darling Basin Authority to get cultural water, and next they're shouting "burn the school". Flaming radicals,' said William.

Arana stood before him, arms folded.

'Fuck the parliament, let it burn,' said Jack.

'Yeah Unc. Most of you oldies just sit there. Yes, boss. No, boss,' said Arana.

'Don't cheek your father!' Merle raised her head from the chair with a killer look.

'The farmers are clear-felling native landscape, killing the wildlife,' said June. 'It's tax deductible, crying for government handouts. It's not Murries draining the coffers, it's whinging farmers.'

William poured himself a glass of water from the tap. 'Bundi reckons if you drink water from a tap in Brewarrina, it's stinking bore water.'

There was a soft scratching at the door and Duckie walked in looking like the living dead. Her face was pale, and her hair hung around her young face like a tangled mop. She chewed her nails and sat on the edge of the lounge.

'Aunty, sorry to intrude, but the grapevine says you know what happened to my man.'

'I'm just asking a few questions,' said June. 'Did gunjie interview you?'

'Yep, just told me to keep quiet and stay home because welfare might come for our son, but. Shithead holes.' Duckie picked at scabs on her arm and sniffed as June patted her back.

'It's all right, sis. When did you last see Thommo?'

'He gave the little fella a bath and played with him in the playhouse he built.' Duckie sobbed and held a tissue to her nose. 'He loves his son so much, but. He cried when Billy was born. Like he was going to change the world so the next generation could live a better life, but with a pure running river. He swore he would beat the river thieves. When I find him, but, I'm gunna to tell him about how I nearly got raped and he will probably kill that bikie or something. Billy keeps waiting by the window and calling for "dada". He took his car out to the rocky highway to meet someone. But people say you might know something. Why would they say that?'

'I don't really know what happened to him. But I suspect he's dead, I'm sorry.'

'I bet it was a bad drug fella who wanted something. I don't know about nothing except Thommo never ever did that before. Maybe he just got done for no reason, you know, but, a serial killer,' said Duckie.

'I'm going to try to investigate the reasons. Then we can inform police and get justice.'

Duckie nodded, then like a wraith she drifted towards the door, then stared at June. 'Digging up stuff, who cares, but, about that kind of trouble anyway?'

'I do,' said June, and she wrote in her notebook and sighed.

* * *

The next afternoon the family argued in the kitchen, while Jack quivered on the lounge with a pillow over his forehead. June tried to keep the peace.

'You just don't want me to be with Steve and save the planet,' said Arana.

'Choose anyone you like,' said June, and stroked her cheek.

'Find a quiet fella at school. Someone your own age, not some uptown big mouth,' said William.

'Not interested.'

'I'll tell you one thing, he better not take advantage of you, or he'll be gone.'

'Is that all you think about? He's coming to the ball,' said Arana. 'You don't understand, young people are respectful towards each other, whether we choose same-sex partners or people of different ethnic backgrounds. We respect trans and gender fluidity. I have great Asian friends; two girls at school wear the headscarf; I stick up for them when racists yell out stuff.'

'Oh, so now you're telling me you got a lesbian Muslim group of mates?' laughed William.

'Two different groups.'

June nodded and gave a determined look at William. 'Now, brother.'

'It's not okay for some older bloke seducing a young innocent girl.'

'Not so innocent, Dad.'

'Don't tell him that, dort,' said June.

William huffed, while June jumped up to sweep the floor.

'Eh, don't throw those butts out,' said Jack as he gathered the butts from the dust pile.

'Jack's into recycling.'

June put his butts in a plastic bag.

Jack stood up. 'I keep quiet these days. They locked me in the jillawa next to stinking pans when we was kids at Kinchela. Ate bread and jam. Froze me mundowies off, bringing in the cows. They whipped us with a stock whip.'

'Okay, Unc,' said June.

'June pays the rent, you and I only boarders here,' muttered William.

There was a knock at the door.

'Who that be?' said Jack.

'I told you, Steve's coming,' said Arana.

The young, handsome, dark Aboriginal man stood at the open door. June welcomed in Steven Allunga from Sydney. She knew already he had a shy disposition but an assertive stance, and she liked him. She sensed nothing would ever be the same from that moment on.

'Jeez, I thought I wouldn't find your house. Google Maps to the rescue,' said Steve. He smiled at Arana with love as she clicked his photo with her phone then folded her body against his. He leant forward and shook the men's hands. William cracked his knuckles. He twisted the signet ring on his finger back and forth. June reached out and patted her brother.

'Where'd you say your mob come from?' said Jack.

'Up north.'

June observed the shy young man gently take their measure as he lowered his head around Elders. He turned off his phone and put it in his pocket. She thought that a lot of the young fellas might learn from him. The strong, high-cheekboned face was open and his eyes shone under eyelash fringes. He had a magnetic presence, and June felt a shot of panic.

'I was working cattle up there ... Oh, must be twenty years back,' said Jack.

But William launched straight in. 'Reckon you're a bit of an environmental radical, eh?'

Steve looked amused, but out of deference to an older man he lowered his eyes again. 'Nah, I write some articles and talk at a few meetings. Tell people about their legal rights to water and land. I'm studying at New South Wales uni. I liked the look of your investigating service, Aunt.'

'I'm going to help her,' said Arana. She glowed, carrying her identity like a torch. Her hand reached out and grasped his, and he hid it behind his back.

'We can only try to take care of the earth. We're waking up,' June said.

'He's studying law,' said Arana.

'I did fine without a uni degree,' said William.

'You're a council worker, Dad.'

June saw a wedding, and the young ones in graduation gowns and red, yellow and black sashes.

'Watch out, they'll have the leg irons out. Big edumucated fella. Them lawyers make a bit, eh? I'll be round for a loan as soon as you're graduated.' Jack rubbed his fingers together.

'You dragging my daughter off to demonstrations?' William stood up eye to eye with Steve.

'You gotta be careful, dort,' said Merle. 'They'll video you, and you get a file, marked for life.'

'That's why I didn't go,' said Jack.

'You're just a jarjum,' William said. 'Listen! You do the right thing, they'll leave you alone.' His finger pointed at Steve.

'Sorry you don't agree, but we respect you,' said Steve. 'We got to fight police harassment. When I first went to Sydney I was picked up by police at Central Station. They bashed me with a telephone book and threw me in the cells. I hate being locked up.'

'We all been in a lock-up,' said William.

June saw Steve's courage and fire and was spellbound. Maybe he could help her find out about Thommo. He was training in law. This might be a breakthrough for her investigation.

'The gunjies usually give you a fair go,' William snarled. June pushed him back in his chair.

'Can't stop the gunjies,' said Jack. 'Might be spying on me in the jillawa.'

'It's them two local cops I'm worried about,' said William. 'How about that missing fella? No real investigation. Smells of corruption.'

'He's not going to just sit back and accept injustice,' Arana said. 'We've got to resist the dominant ideology.'

'You swallow a dictionary, girl?' said June.

Everyone was quiet except for the sound of a clock ticking. June felt her heart thumping and held her chest.

'That's right, young ones have got to speak out,' said Merle as she made a cup of tea, tipped Tim Tam biscuits onto a plate and laid it in front of Steve.

Steve checked his phone. 'I'd better go to training.'

June's eyes were on him as he left. She pressed her chest to breathe.

CHAPTER 14

12th March

The family drove back early from the corroboree festival because they ran out of food. June had watched Arana dance and was very proud. William had performed a bit of shake-a-leg with ochre all over his body. His red undies had looked a bit baggy, but he was happy.

Outside their house, young fellas were working for the council. Murries in tight stubbie shorts and no shirts – watch out! June had her eye on one of them.

She cooked a dinner of sweet curry with mashed potato while Arana and Merle read books. Back on the lounge, June watched Jack concentrating, his fingers waving mysteriously. He conjured spirits and visions, was mesmerising and hilarious.

'I saw a little one too. Brown jacks, we call them,' said Jack. 'Agabinya wandabaa, and he was brown and hairy, two and a half feet tall. Little tiny feet, crossed teeth, red eyes and a terrible hairy face. I saw him up in the snow at Barrington Tops. We saw little tracks. Looked around and he's gorn, a buuli wind.'

'You always seeing things, old fella,' said June.

'Eh, I'm certain when I'm seeing real things and when I'm not.'

'He says he seen Dooliga, yowie, the Hairyman. Eh, brud? You sense him and sniff him. What a stink, eh?' said William.

'I seen him.'

'You seen him, Arana? The Hairyman?'

'No, I haven't. It gives me the shivers. True, eh, Uncle Jack? Sometimes I sense something's behind me. If I spin around quickly: just a shadow.'

June hushed her, and they spoke in soft tones and tried not to think about those hairy men who followed you; they were wandabaa. Yet they were playful and sneaky and a menace.

'You take care now, eh, Arana?'

Arana kept checking her phone, and after a few chimes she grinned and cuddled up to old Jack on the couch.

June plonked down the plate of hot food and cut the steaming damper.

The family sat down to dinner, but there was a loud whistle outside and William's head jerked up.

June looked at Arana and knew how in love she was. 'Who's that?' she asked.

'That's Steve.'

'What's he want?'

'Nothing, he's come for dinner.'

'Is he bringing the gunjies with him, or they callin' round later?' William grimaced and brought another chair to make a place.

'Dad, please be nice, he's nervous enough, it's a big footy game.'

'He's been asking questions round town, even to coppers, he's a big stirrer,' said William.

June opened the door and Steve arrived; there he was in a nice clean white shirt.

'Eh, here he is, look out.' June looked Steve up and down and muttered in a low voice to Arana. 'You gonna wiggle your moot tonight, or what?'

'Hungry, Steve?' asked Merle. 'Grab yourself a feed. Help yourself. Curry and mash.'

Steve joined everybody at the table and ate quietly.

'You're from up north, Steve?'

'That's right, but I attended boarding school in Brisbane. I came to Sydney for uni,' said Steve.

'Too hot for you up there?' said Jack.

'Have a beer?' said June.

'No, thanks, don't drink.'

Steve pushed his plate away and stood up to gather the empty plates and take them to the sink.

'Well, maybe you should start,' said Jack.

June had observed Steve place his knife and fork together on the plate. Trained in a foster home – she knew those English manners. At school she had to chew with her lips pressed close and elbows squeezed by her sides, waiting for the approving smile from the house mistress.

'You like curried goolbree … emu? We had a roo, but that mob up the mission took it,' said William.

'You know anything about Duckie's missing man?' Steve asked.

The room hushed and June stood, ready to take notes. 'Only someone wanted him gone, eh?' she said.

Steve wiped the sink and Arana glowed.

'How about the Knockout?'

William began a war cry, and the rest joined in. 'Narcoo Warriors! Up the Warriors! The Warriors!'

'That young fella had rubber legs, he was jamming the yarndi all night,' said Jack.

William doubled up with laughter. 'He'd been eating emu! He didn't find which way to run!'

'They smoke a bit of yarndi, do they? The kids up here? Get it from bikies?' said Steve with a hushed voice.

'They knock it off from some big plantations up in the hills. Everyone reckons that the gunjies are running it, and that Mr Wright is involved,' said Arana.

Everyone went silent and June put her hands on her hips. 'Arana! Just keep that talk to yourself,' she said.

Steve spoke out. 'No one in town was speaking about that Thommo disappearing because he interfered with water stealing by the Big Cotton farmers. Somebody must've known something, like you, Aunty June. People are too scared to oppose the corrupt bastards. Drugged or drunk, doesn't matter to the government – keeps the people submissive. Yarndi's just another tool of oppression, and we've gotta fight back.'

'Black against white?' said Jack.

'Consider racial domination and subordination, it's a political construct, this whiteness,' said Steve. 'You won't get killed by the police just because you are a white person, but someone might kill a Blackfella to shut him up, eh, Aunt? If you're planning to really upset the big end of town. Just like that Thommo. Or if you ask too many questions.'

'What? Pass the bread,' said William. 'You got a number of corrugated iron words there, brud.'

'That's dangerous talk, fella,' June whispered. 'It can lead to terrible trouble for us.'

Jack put a CD on the player and Charlie Pride poured into the room.

William sat next to Steve. 'Better you keep quiet,' he said. 'We try to keep good relations with whites.'

'I'm sorry if I offended anyone.' Steve looked at June and she shook her head, thinking that he was saying too much.

'That Thommo fella, our deadly cousin, we reckon he's gone. He's got a family round here, they don't want to be hurt, but we want to find out what happened,' said June, and she looked agitated and anxious.

'We can't just lie down and take it – Murries have to speak out,' asserted Steve.

'We make enough noise to save the dying river. We plan to agitate when it means something,' said June. 'But it's time for young ones to step up. I know I talk too much. I need to listen more. Even to drought-stricken farmers. Poor things. We might take up a collection and buy bales of hay to feed the animals.'

'Not likely,' said William.

Steve stood up and turned to thank June. She watched him, thinking how the lad had been well brought up.

'Well, that was a great dinner,' he said. 'Thank you very much. I have to eat and run. I've got business to attend to. Then the coach wants us up the oval for training.'

William opened the door, and June saw he was happy to see the outspoken young man leaving.

'You'll need all the training you can get against our team,' said William.

Arana walked Steve out the door and looked lovingly at him. She held his hand despite the hard looks from her father. The young ones strolled to the gate.

Through the window, June saw them kiss passionately, then Steve pulled away and jogged into the dark street. She drew the curtains shut.

Arana shut the gate behind her and leant against it. She looked up at the stars and sighed. Night birds fluttering overhead. She came back inside and sat.

'He talks same as Gary Foley,' Jack said.

'Lovely bum, like Adam Goodes.' June laughed.

'What would he know? He's not even drinking,' Jack said.

'What business does he mean, love?' June asked Arana in a whisper.

'Like killing environmental activists.'

June rushed outside, leaving the family wondering.

'I'm gunna check the dogs, that big red one will eat him,' June called out and hurried across the street to catch up with Steve. Something told her that he was trouble. As if a curlew was crying out or a willy wagtail was singing. 'Eh, hold on!'

Steve looked over his shoulder and June caught up.

'Everyone's been talking about you,' she said. 'Murri grapevine's lightning. You been asking questions about that fella?'

'I read info on his death on social media,' said Steve.

'Yeah, people are frightened. Don't say too much around town! He might have sabotaged the cotton farmers' pumps. He was wild, all right,' said June.

'I got permission from his people and the Land Council,' said Steve. 'Who do you reckon killed him?'

June took a grip of his arm and pulled him close to whisper. He might think she was mad, and too protective, same as every Aboriginal aunt or uncle who tried to tell him what to do. She saw he was alive with energy to change the world.

'He didn't go walkabout,' she said. 'Some relations in Sydney or anywhere would've told his family where he was. And I took a photograph of evidence in to the police, but they didn't follow up on it. A boot print. I'm sure it was the same biker boot as Bam Bam or one of his mates has.'

It startled Steve; his breathing was heavy.

'You saw it yourself, eh?'

'That print is evidence. We went out to the last place I knew he went,' whispered June. 'Those police, they don't listen to us. Told us they found nothing, no body. They said the boot print was not part of the evidence.

'It might be on file.'

'That Thommo was only thirty, got a kid in town. I reckon he's buried not too far. The gunjie said it was a drug deal and maybe he ran off and he sold yarndi, but he's not like those bikie blokes, selling heroin to kids on the mission.'

'Who are those whitefellas? What names?'

'Elders wanted to pull out the nulla nullas and clout them. It's frightening what's happening up here,' said June. 'We've done a bunch of work building that bridge between white and Black, but you listen to me, Steve – what I've told you, keep my family out of it. Leave Arana out of it. I've already had blokes sitting in cars outside my place. I don't like having to deal with guns. Okay? Now go on, get going.'

June walked back inside as Arana took her jacket and stood at the door. She smiled at the family.

'I don't think you should go out again, darling girl. I don't want you chasing that fella around town.'

'I'm seventeen now. We only go for a walk, don't worry about me. He's a strong fella, and I have been going with him for a few months.'

Merle looked shocked, but couldn't say anything, and just shook her head at William and laughed. 'Months? You must be joking! We never even guessed,' said Merle.

CHAPTER 15

17th March

Early on National Closing the Gap Day, the morning was cool, summer was over. Wayne Johnson arrived at June's house to meet William and Jack and go hunting kangaroo. He sat on a chair and looked uncomfortable in the kitchen while June worked on her computer. She stood up and made him a cup of tea.

'You want milk and sugar?' said June.

'Thank you so much, Mrs Thomas. I hear you opened a detective agency, impressive.'

'Call her Aunty, everyone else does,' said William.

'Got some good cases?' said June.

'Lots of unsolved break and enters. We try our best, but you might have better insider information. Some of your cousins down the pub were teaching me some Aboriginal language. Gamilaraay it was. I know dinewan – that's an emu. And if you want a drink, you ask for a ngaru-gi, but that's hard to get my tongue around,' said Johnson. 'Oh, and they said I should ask Aunty June for a doori.'

They all stared at him.

'That's some kind of meat, isn't it?' said Johnson, and June blushed.

William roared with laughter and Jack whistled. 'Nah, that'd be thii.'

'He might get hit over the head with a rolling pin, eh, June?' said William.

'I usually like a young one, but I might have to draw the line at a gubbah gunjie,' laughed June.

Wayne looked mortified. 'Those cousins tricked me, eh?' he said. 'I'm sorry, Aunty. It's disrespectful.'

'You can go out and get me a nice leg of bandaar – some roo. And get me a photocopy of the missing person report.' June smiled.

'Sorry, Aunty, I'm not allowed to do something like that, not for anyone.'

A bit later, they all met up in the bush near a wide plain where emus and kangaroos grazed at dawn, with the sky pink and a mist rising from eucalypt trees. Johnson had a hold of his excited dog on a chain. It leant against his master's leg.

Arana drove up in the Holden, and Jack, June and William got out. Soft sunlight was coming up and faces glowed. June breathed in the scent of the bush and felt alive. She observed that Johnson noticed Arana's beauty, and Arana waved with a shy smile before flinging her head sideways in a dismissive action and taking out her phone and texting someone. Then she drove off over the bumpy track.

'I saw a young woman dropping you off. Good looker. Who is she?' said Johnson.

'Don't you worry about her. That's my daughter, and I can see what you're thinking. You come from a decent family and you're not a racist, so it's not personal, but son, she's taken.'

'Just curious. She's really beautiful. With that long shining hair, she could be a model.'

'So everyone keeps reminding her. She'll get a swelled head,' said William.

'Who's she going out with?' mumbled Johnson.

They held up the weapons and checked the barrels for dust. Jack held the ammunition with great care as William loaded the guns. The bags for animal bodies were piled next to the car. A cool breeze blew up, and they held their shirts close to their throats.

'She's got a uni fella,' said June. They all shuffled their feet in the dust, and Johnson gave a deep sigh and he shrugged his shoulders.

William changed the subject: 'Her fella's with the Greenie movement. Might dig up the irrigation channels when no one's looking. The young people are wild about the river water being stolen.'

'That would be breaking the law. Destruction of private property. Cotton farmers would be within their rights to defend their land.'

'Our land. Our river too.' June held her annoyance and tried to be nice. 'We need to find out the truth, eh?'

'You got your rifle?' said Johnson.

'I'm ready.' Jack tucked his gun under his arm, and June thought how it was too dangerous to give Jack a gun.

* * *

The men walked through a patch of gidgee scrub, looking for young bucks or does without joeys, and they moved as quietly as possible. Wayne Johnson was excited, and he watched the figures ahead, shadows against the rising sun, the light playing on their upturned faces.

'What are we looking for?' said Johnson.

'We aim for a head shot, then the dogs bring it down,' whispered William. 'Then we kill them. There's a danger of a kick from strong hind legs. Rip your guts out.'

'I'll watch you.'

They hunched over and then crawled through the grass and stayed still for a long time, watching for kangaroos.

Johnson thought about Arana's long legs running in the wind and how the tendrils of curls flickered around her golden face. He had seen her looking at him; yes, she had stared at him, and he was certain there had been something. The tendrils were in her mouth as her pink tongue had licked them up. Windy breeze flew around her as she had helped Uncle Jack out of the car. A Holden. Old style. Nice. Her black lashes as thick as cat fur. How would they have felt on his cheek? Those long legs going all the way up to heaven. No, she was William's precious girl. He mustn't think of her like that. Out of bounds. But she had looked at him as she drove off. She had her eyes lingering on his. He didn't imagine it. Frisson, that's what it was called. Hair blowing out the side window. Impossibly long and made of dark silk. His thighs ached just thinking about it.

William took his 202 rifle and loaded it, ready for the hunt. Johnson saw how he was careful with it. And he had a licence. Responsible, like he had learnt at the academy.

They walked quietly into the mist. The hunting dogs sniffed the earth, catching the scent of roo. Jack was slinking down low, and Johnson saw he was surprisingly swift on his feet. 'Nimble Jack' they used to call him. The long grass was high enough to shield them. William motioned to Johnson to get down and hush.

He remembered her dark eyes and thick lashes, her eyes on his, in his, just for a moment. Was she a flirt? What would it take to win a girl like that? When she turned to get into the car, her dress had ridden up, revealing golden thighs. It would be heavenly to touch that skin, running his hand down her back. See her naked. He shook his head, with William beckoning him on. Down on the earth smelling the rich aroma of dust and roo dung. The men surrounded the kangaroos in a gully. Jack was the master of surprise; his instinct was his old knowledge.

Then the dogs were at the kangaroos. A loud bark and yelp – yes, they had a young one down. The dogs were so fast, and then Jack was with them, calling them off and holding them back with gruff commands. William finished the roo with one shot, then the men lifted the dead animal onto Wayne's shoulders. He was the youngest – he could carry it to the car. He was proud, now, of this man's work, and of being out there in the semi desert with Aboriginal men. It felt great and ancient, like he was one of them, and on his way to initiation.

They pushed the roo into the boot of the Holden, but the legs and tail stuck out, so William tied the boot shut with a rope while the tail flopped outside the metal hood.

'Hope a gunjie doesn't pick us up for defects. Eh, Wayne? I got no tail-light on the right rear,' said William. 'Those bike clubs grow dope out here, I heard about it but couldn't find anything.'

Johnson shrugged.

'Duckie lived with a bikie, that Bam Bam,' said William. 'When she was too young. Illegal, I reckon. They had a farm. June saw her when she was eighteen, all bashed up, black eyes, busted collar-bone. She said she fell over the dog bowls. But we suspected Bam Bam. He was drunk.'

'Should have reported it to police, domestic violence,' Johnson said and shook his head. He was disturbed – this kind of stuff was upsetting.

'Nah, she should have packed her stuff and run for her life.' William stood with arms crossed. 'That man had a reputation. He always gets off. He has a really good barrister. Dealing ice is good money – he's loaded.'

Johnson listened and mentally noted that Mr Bam Bam might be on a list of offenders somewhere.

'Don't talk about it. You get the tail, all right, eh?' said Jack.

'June will be wanting to make soup with that kangaroo tail. We better stop at the shops and get some bacon and onions. Tastes really good like that.'

'Sounds delicious,' said Johnson. 'I've only ever eaten roo from the supermarket, marinated. Say, William, could I ask Arana out?'

'Nah.'

'He's real edumacated,' laughed Jack.

'I might study that too, one day.' Johnson looked embarrassed and wished he had kept quiet.

There was an uneasy silence and they all looked at the kangaroo.

'Me too, brud, me too,' said Jack, and he slapped Wayne on the back, and they all laughed.

CHAPTER 16

25th April, Anzac Day

They bustled around the house then drove downtown to plaster their posters for the environmental meeting around the town. But they waited with respect until the Anzac march was over. June had held up her Aboriginal flag and clapped the men and women marching past. She refused to wave an Australian flag, with its bloody British one in the corner. *Britain by night*, she thought.

The pubs were full and not many people on the street. The sound of games of two-up echoed across the road. June unrolled the posters. She handed Merle the homemade flour glue and a brush. They brushed the back of a poster and stuck it over the advertisement: 'Eat beef you bastards.' They continued up the street, pasting posters on disused shopfronts. There were dozens of them. The town was dying. The pineapple-fruit can of glue wobbled on the cracked pavement, and June wiped the white droplets from her dress.

'Looks sus,' said June. They laughed and dribbled glue across the paper.

They were like guerrilla warriors. No one was around the dusty streets to make stupid comments. June led the way to the levee bank to rest, and they sat on milk crates among the plastic bags and aluminium cans. A few skinks scuttled off and the galahs above never stopped squawking.

Life might be a long pathway weaving like a snake. June looked at her hands and turned them, wondering about the songlines of the rivers. The river Darling was their lifeline along its winding path. Rivers of stars in the Milky Way, the whispered Madukka and paths to follow for the Dreaming stories. Connected like a spider web to every living thing. June liked to sit by the running river and be quiet, just listening to nature. The ancestors watched for blossoms on a certain white tree to tell them it was time to harvest water lilybulbs to roast on coals and eat. The river had brimmed with native fish, but the carp had eaten everything and they were inedible.

'Look at that green puddle,' said Merle. 'Drinking water costs six dollars a litre at the IGA supermarket.'

June's vision of destruction was terrifying: the river red gums would die in the next ten years because of the sucking up of their river. The northern New South Wales and Queensland cotton and rice farmers took the life from the precious stream. She could see the growing salt bush taking over the flood plains and marshes. The kurrawin, great ancient red gum trees, had collapsed, and the sentinels clogged the shallow pools. They had seen it all, the giant pools of irrigation on the huge stations with the evaporating water turning to mist. Economics would force them to watch the precious river in its destruction. River ghosts and piled tree trunks tumbled in black masses along the dry bed. The Murray–Darling Basin would cease to be full of fresh water.

Old Mary from the Reserve was sitting nearby on a blanket under a prickle bush, sewing. She had a Gobblers' vest in her hand and was embroidering a bush turkey. She tucked it under her skirt as June approached.

'You catch any local fish, Aunty?' said June.

'All gone now, I reckon,' Mary said.

'There are no Murray cod left and they might even be extinct!' June said, and nodded.

'I used to fish every day, but no good.'

'The gums are bare of leaves, no canopy, eh?'

'That right, no maluman, manna sugar to eat,' said old Mary. 'Make you cry, all right.'

June had met the Aboriginal Environment Group to talk about human greed and poor decisions based on individual farming needs. Big farmers had diverted over a thousand kilometres of the river to irrigate the enormous cotton farms. Cotton chipping was the work that Aboriginal people had carried out. The work they did to survive. They were like the African slaves in the olden days of the United States of America except the Murries got paid basic wages. Just because some local cotton farmers supplied a few Aboriginal screen-printing fabric enterprises, Big Cotton thought that the Aboriginal people were in their pockets.

'I reckon the manufacturers should just buy cotton from countries like India and Egypt who need the income. Not from Australia,' said June.

'Reckon.'

'They bloody well fixed the meters to give false readings. What a bunch of crooks!' June said. She no longer believed in the farmers' ability to comply with laws.

'It's corruption of public officials; they actively stop investigation into water theft, we all know that,' said Merle.

'We should run them out of their well-paid jobs and put them on the unemployment benefit. See how they like it.'

'You do that, girl.'

Merle gave old Mary ten dollars and a hug while June helped her roll up her sewing kit.

'You want a lift to the Reserve?' June asked.

'I gotta go hospital see doctor about me foot, but bit later.'

'Are you sure you don't know what happened to young Thommo?' June said his name out loud to the air, to the birds and bush.

'Nothing to do with me. He broke our law. His wife hit him once with a frying pan. Too cruel, that one.'

June and Merle left her on the riverbank and walked back to their car. They piled the leftover posters on the back seat and drove home. As they turned a corner, a big motorbike passed them, and June recognised Bam Bam heading to the riverbank followed by a motley collection of bikies. The roar of the bikes sounded like lions on the hunt.

* * *

In the afternoon, June looked out her office window and saw that their environmental signs were now graffitied with obscenities. Her window had also been spray-painted with abuse.

William arrived, and they pulled down the handwritten sign outside the Rose Café because it named her and Merle as Black stirrers and other more disgusting things to do with their vaginas. She didn't think hers was that big.

A dark new SUV pulled up. It had tinted windows and didn't appear local. Three young men in black t-shirts stepped out onto the dusty street and lit cigarettes. William watched the car, with its hotted-up engine. June pulled at William's sleeveless shirt. The flannelette was sticky with sweat.

'What are you up to, fellas?' asked William. 'That you in the back, Cowboy?'

Cowboy stepped out of the car.

'They're on their way across from Adelaide. They're Socialist Nationalist Resistance. They want a pure white Australia,' said Cowboy.

The largest man unfurled a poster about the fascist movement.

William laughed. 'So we're not good enough, eh? Too black, are we? You look like you're a bit of a Mediterranean yourself.'

'You like birdwatching, don't you – or was that someone else the other day?' June asked.

'We're being overrun by gays, Asians, Muslims – it's an Islamist conspiracy! It's a massive change. We're stopping foreigners, we don't want them.'

'Yeah, in this town you'll find Hussein from Turkey, but he's pretty quiet,' said William. 'His wife makes the baklava pastries for the Rose Café. They're pretty tasty. Oh, June, maybe he's talking about that old Afghan bloke buried in the Bourke cemetery whose dad drove the camels from Adelaide.'

'It's Australia first!' Cowboy said. 'These fellas are driving through Broken Hill, Wilcannia, Walgett, spreading the word all the way to Toowoomba to stop the mosque being built. We are patriots!' He stood tall next to the visitors.

June whispered that they should go. Her hand clung to William's shirt. She could sense the hostility and it was very frightening. Her anxiety pulsed.

'Eh, fellas, we don't need that rubbish in this town,' William said. 'Many of our Koori brothers are Muslims, like Choc Mundine. So why don't you take your poster and piss off.'

One man with a shaved head stepped closer to William and hissed as June stuck out her chest next to him. She looked at the young man's t-shirt, printed with a white skull. The other men stood in a clump and jeered at June as she turned her back and put up her sign with drawing pins. One man stood near to her poster and sniffed it like a dog.

'Let's piss on it ... White revolution is on its way, the only solution,' he said.

'It's all right to be white!' said Cowboy as he tore down June's poster. He handed it to her and smoothed the ripped paper, and she shivered with the closeness of his stinking cigarette breath. 'Sorry, don't know my own strength.'

Another of the men hammered in their own swastika poster where June's had sat. Her fists squeezed tight, and she wanted to punch them.

'You touch this sign, and we will come lookin for youse.'

'I'm not afraid of racists,' said June as she looked at the tattoos of Nazi symbols on Cowboy's biceps.

'You should be!' he snarled. 'Time to call out the Black scourge and sodomite filth and Chinese and Muslims.' He smiled and patted June on the head.

'Aussie Patriots!' said one as he pressed his thumb to June's chest.

Cowboy pulled the men away from June. 'Don't touch her, leave it,' said Cowboy.

June looked at Cowboy. What was he saying? Protecting her? It seemed absurd, but she smiled at him and pulled at her brother's arm.

She looked at their black clothes and military stances. They were young white males who probably grew up in nearby areas. The street was empty, and June felt vulnerable as the men stared and growled. But Cowboy gave a thumbs-up. William looked back, and she took hold of him and dragged him to their car, and they drove off. Sweat poured down her back, and June sighed with relief.

'I'll tear it down when they leave,' said William.

CHAPTER 17

30th April

It was autumn and getting colder at night, you needed a blanket. June could not bring herself to go and ask for any more news about Thommo from police. She had a fear of even going past the station. An image of being arrested herself made her quake. She was terrified of being set up.

Then they heard that a man carrying a knife had been arrested in Wilga Primary School grounds. He had waited until the students gathered after the raising of the Aboriginal flag for Reconciliation Day. He said he was looking for a teacher called Merle. He was released without charge. He said he was just cleaning his nails with the knife.

Merle and June had joined the local community Inlands River Network. They looked at the breeding habitats of birds that had been destroyed. The government had drained the Menindee

Lakes, and there was no more flooding to bring life to the marshes for bird breeding. June mourned the fact that there were no more frogs in the wetlands. It was an immense death of many bird species due to climate change and water thievery. The Darling River was now a trickle of yellow mud.

The women walked along the river with the bush care group, made up mainly of old whitefellas and Elders. Up front, girls and boys from the local school leading the animal tracking. They went out nature-spotting regent parrots with binoculars and carried out surveys with the National Parks and wildlife mob. They were looking for the hairy-nosed wombat, as it was nearly extinct, like the southern bell frog.

Bitterns and painted snipe, plains wanderers, Mallee fowl, emu wren were all in decline. The magnificent magpie goose was already extinct in that area, as were the black bittern and the squatter pigeon. June wrote out the list which included the endangered section: red-tailed black cockatoo, regent honeyeater and the western whipbird.

It was early morning in the national park. The track passed the river's steep grey walls, carved with crevices from rain. About twenty people stood by the meeting place. Cockatoos screeched and crows pecked at something dead down a dry rivulet. White bones poked through the cracked mud.

The group gathered in a clearing. They slapped on sunscreen and adjusted sun hats.

'It's not bloody dead,' said an old bearded Elder. 'This mighty river is waiting for us to help it. I had a lot of hate in my heart against whitefella. A fella without Dreaming, without Djugabarra, he will have to search for it. We survive, we are not going anywhere. We should challenge the government. Can't go without water. That serpent will wake up and be real angry, bring destruction.' The old man sat down and adjusted his leather bush hat, and smoothed the eagle feathers and goanna claws down on the brim.

Effie the Sydney environmentalist stood on a little hill under an ancient coolabah tree. June was half in love with her already. Fiftyish with a bright face, long dark hair and no makeup, just vegetarian glowing health and wonderful breasts in a tight safari shirt. Maybe June was turning lesbian. Effie grinned at her and licked her lips. June liked the appearance of a nice breast. A nice nyamu.

'I wish to acknowledge the traditional owners of this land, and pay respects to Elders past and present,' Effie said, and nodded at the older women and bowed her head. 'You will already know about the western cotton farmer. We're prosecuting him. He illegally extracted oceans of water over two years. But we are also under attack with legal action brought by Warren Wright. Our local investigator June has evidence, but the police have done nothing.'

'That sounds right,' said June.

'The deputy director general of New South Wales Water met with irrigators and gave confidential information to guide their lobbying of government. They have referred it to ICAC. Water New South Wales says it cannot make a comment,' said Effie.

'It's corruption!' said a young man with dreadlocks and nose rings. 'One of our protesters has already been murdered by irrigators!'

'It is a failure of the compliance arm of government to enforce water regulations.' Effie paused and tilted her head back to drink from a reusable water bottle. 'We were followed by a truck with "W. Wright" written on the side. They tailgated us for an hour. Tried to run us off the road.'

June stared at her because she was not sure Effie knew who they were. Did she realise that her Murri family was at risk from dangerous men? That sleeping at night was tainted by constant fear of intruders at June's house wanting her to stop inquiries? Big agri business was just one of the enemies.

'That's their tactics. Among us are traditional owners. It's our water too,' June said, and she fluttered her eyelashes.

'And two motorbikes with men in bikie colours terrorised us – frightening.'

'Shitheads.'

'The irrigators emptied a fifth of the Macquarie Marshes,' said Effie. 'Now they are drying up, with no wetland for birds. We're facing complete river system collapse and extinction of species.'

'It's the end of the world,' said June.

At this moment a flock of galahs arrived overhead and settled in the coolabah trees. They ate the seed pods and threw sticks down on the people.

June raised her hand and Effie nodded. 'We Aboriginal people are anxious about the dying river and the spirit Moodagutta and all dying species. My favourite is the Murray–Darling carpet python,' June said, and she looked around as if a python would appear. 'They have beautiful, patchy dark skin and they have been disappearing for years. I remember seeing heaps out bush when we were kids. We even kept them as pets. Found them near Mount Gundabooka, our sacred site.'

Effie smiled, and they all wrote notes. 'So, what do we know?' Effie said. 'Overgrazing, salination, weed invasion – the environment is inhospitable for birds and animals. We lose habitat like hollow trees where nesting happens for cockatoos, parrots. The bustards used to be a common sight but now we hardly ever see them,' Effie said.

'Let alone shoot one for dinner,' said Merle, while June nudged her in embarrassment.

They walked along the dry creek beds and looked for species to record. In the bushes they saw a small clearing and a man in a camouflage jacket, who sat on a high rock, holding binoculars, facing them. Cowboy's patch of marijuana gleamed from the undergrowth in scintillating green. A black sign with skull and crossbones hung from a tree. Then the bikie himself appeared, leapt down from the rock and ambled over to the bushwalkers. He wore tight white underpants and tattoos. He held a machete. June was worried.

'Gidday, fellow bird twitchers! I thought I spotted the rare double-breasted booby, but it was this lady here.' He pointed at Effie's breasts and smoothed his white beard, which had wildflowers in it.

'Just enjoying the wildlife, eh, mate?' June grimaced at him and pretended she couldn't see the crop.

'Yeah, camping – you won't tell on me, will you, Aunty?' Cowboy scratched his underpants and gazed at Effie's breasts again.

'Nice boobs, love,' he said.

June motioned to everyone to get going. 'You're okay. We're heading up that hill to look for species.'

The bush-care group hurried off and didn't look back. Everyone had some knowledge of Cowboy and the rumour that he collected guns, including an automatic rifle. June looked back at him, and he blew a kiss. She walked quickly up to Effie. 'Where are you from? You speak with an accent,' said June.

'I was born in Europe. I'm Greek.'

'I sometimes think I might just move to England,' said June. 'Where one of my grandfathers came from, down in Eastbourne. Entitled, as I am, to citizenship. I could cover up my brown skin with makeup and grow pink duchess roses, visit the *Antiques Roadshow*, that'd be nice and safe.'

Effie doubled over with laughter. She winked at June. Then they stopped still, the roar of motorbikes in the distance.

Walking back to the vehicles, they were met with the unwanted vision of graffiti all over their cars: 'Piss off Greenies, we'll kill you!'

CHAPTER 18

12th May

On the weekend, after a community barbeque, the family went out bush, and young Wayne Johnson pulled up beside them in his ute. William had an idea to take the young fella and show him a carved tree and some special rock painting sites. Circles in ochre told a story of sacred places and ceremony. Teach him respect. The sky was bright blue and full of cockatoos and galahs, with white gum trees bending towards the riverbank.

Arana was wearing short shorts, and lilies perfumed her body, and she looked so beautiful that young Wayne Johnson couldn't stop staring at her. He kept looking back at her and offering a hand to help her over logs, but she laughed and rolled her big brown eyes at him.

'Nice old Holden, Will,' said Johnson.

'Yep ... This is where we hunted emus,' said William.

'I am thinking of going vegan,' said Johnson.

'Me too,' said Jack. 'They good eating? You gut them before putting them on the fire or what?'

'We made a decoy horn from a piece of hollow coolabah tree, rubbed it down with a stone until it was thin as cardboard,' said William. 'Stop one end up with beefwood gum. You blow across it to attract emus. We set up a cord in a triangle net to catch them. Some fellas lie on their back and do a bicycle movement and those silly emus would come and we'd kill them.'

'I'd like to look at those carved trees,' said Johnson. He really wanted to lie down in the grass with Arana's head in his lap.

'You shouldn't talk too much about them, love,' June said and shook her head.

'We can show you one, but be respectful, it's a spirit place,' said Arana.

'Oh, I will be. Really, it's great that you can show me stuff. I appreciate it, really.'

'Her name, Arana, means magpie, that's Aranda language. That's why she chatters,' said June.

Johnson looked at Arana and blushed. She was a pretty girl, and no red-blooded hetero male could be immune to that. He imagined her in his arms. Her hair was incredible, flowing like dark liquid across her shoulders. He wanted to reach out, take a tiny lock of hair and taste it. His fingers twitched. He watched her long legs as she hopped between grey rocks. Wayne wished he could pick her up and carry her over the rough track.

They walked quietly to a clearing and there stood the magnificent ancient tree, tall with deep carved markings of crosses. Wayne thought he had never before been so peaceful in his life. The family was quiet, crows and cicadas the only sounds. He smiled at Arana but was embarrassed by his eager staring.

William stopped everyone on the track and called out to the spirits to let them know they were coming near and would not hurt the sacred place. Wayne followed William and tiptoed with a big grin on his face. He kept down, and he looked back at Arana.

He gave another ridiculous toothy smile at her as June exchanged shrugs with Merle. They stopped suddenly, and William pointed out another giant white tree. It had a circular pattern, deeply carved. They didn't go too close, and followed William and took his guidance. He seemed to breathe in the spirit, so they copied him. Wayne looked in awe at the mighty living tree. Five hundred years old.

As they walked away, Wayne kept looking back towards the sacred place, unable to speak.

William nodded at June. 'This young fella might be a gunjie, but he wants to learn.'

They sat and drank water on some sandstone rocks. Wayne stretched out and rolled up his sleeves, trying to show off his muscles.

'These carved trees are in northern New South Wales and Queensland down the south-eastern area,' said William. 'Some figures are like men or animals, but here it's all geometry. Lines that indicate a burial. We got to be respectful out here. Spirit place. They might curse people. No mining or cutting down trees. A lot of these trees were carved with metal hatchets, so they'd be about two hundred years. Older ones made by stone hatchet. Bogan and Lachlan River have initiation sites.'

'You initiated, William?' asked Johnson.

'Yep, they took me out bush as a young lad, teach me ngiya law, and I had to survive the desert and make my way home. I only had a bottle of water and no wii, you know, packet of matches. I did it. Nearly died of thirst.'

'The Ngiyampaa mob use songs and ceremony along the Darling River,' said June. 'A special place at Mount Gundabooka, lots of paintings of white nets, fishing dreaming sites. They got that white clay from creek beds and red from ironstone "bombs". We call that place "Kubbur" by that Wongaibon tribe. Those clever men made magic …'

'Some about you know. They can put a magic curse on you, or make a woman come to visit them, true, eh, Merle?' said William as he tapped Merle on the leg.

'That's how you got me? That's nice, isn't it, Wayne? Nah, he's joking. He had to work hard to get me to go out with him. He had a wild reputation,' said Merle.

Johnson swallowed and moved next to Arana. 'Would you go out with me, Arana? I like you,' he whispered. 'Come to dinner at the Chinese, or a pizza? A real slap-up dinner? Even if I am a gunjie?'

William and June roared with laughter.

Johnson's face fell.

'I might one day,' said Arana, and she kissed him on the cheek.

'Those dubai always leadin' us on, eh?' laughed William.

'What did you do in the olden times, Aunty June?'

'We ate lots of bush food, eh, Will? I ate bigibila echidna, baked in an oven at our cousin's, used quills for necklaces,' said June. 'I tried emu, but emu didn't agree with me, but I cook it for them fellas. We still eat lots of kangaroo. Whenever Dad and us kids went fishing, on our way back we would get a kangaroo.'

'I used to collect the cow poo. Burnt it to keep off mozzies,' said Jack.

'You were in the boys home.'

William lit a rollie and shared some fresh damper and butter around, and they ate. Then William told a story. He leant forward and began. 'One night there was a big yellow full moon, and we were sitting fishing for yella belly but catching only carp. Sad, the river so low now.' He used a hushed tone while leaning forward and taking hold of Arana's arm. 'I know a Moodagutta spirit, and it sure lived deep down in rivers, in a hole in the darkest places in the mud, looks a like a long snake. You better not go swimming across the middle because it'll pull you down with the current. Swirl you down and eat you.'

Arana and Wayne laughed. June remembered the visitation of Madukka. The same spirit.

'There're heaps of snakes along our river,' Arana said. 'Brown ones and death adders in the leaves, waiting for you to step on them. They really like biting gunjie. You gotta walk through the bush with a stick and you twirl it in front of you on the ground, flicking it to find Joe Blakes. Or goanna. Tastes like chicken. I can bring you some up to the police station one day, eh?' joked Arana while Wayne nodded vigorously.

William laughed. 'Oh, right. "Hey, Sergeant Blackett, want to have a bite of my goanna leg? You want mayonnaise on that?" I can see that, all right. He's pickin' his teeth with the little claws, nice toothpicks.'

They all rolled around laughing.

'I slept in an Aboriginal cave once,' said Johnson. Everyone became quiet. Now Johnson felt he had something to offer.

They looked with shock at the young man.

'I was up past Tenterfield, out bush with a mate,' said Johnson. 'Near this vast escarpment of stone ridges. Real lovely place. I was with a girl. First love, you know. We found a lot of ancient beautiful paintings in this rock shelter, with ochre circles and some snakes and stuff, a real long way from anywhere. We had sleeping bags, and we camped in there.'

'Bad idea!' said William.

But Johnson kept talking, oblivious to the impact of his words. He wanted to impress Arana. 'Yeah, it was. We made a fire and had a cuppa, and after dark we noticed two big white lights coming up the valley, and we reckoned it was a car, but they were bright and swerving all over the place, but it felt really weird and scary,' said Johnson.

'Min min, you shouldn't have stayed there,' said William. His face was leaning forward, looking intense, and he sweated.

'We thought, jeez, this bloke must stop soon and come up and tell us what he wants,' said Johnson. 'Then the lights stopped in front of the cave, and we were shitting ourselves. It wasn't a car. The lights lit up the whole valley. Creepy. Like flying saucers or whatever.'

Arana stood up with her hands stuck on her hips and stared at him with disgust.

'That might be a burial place. I hope you didn't shout or do anything stupid,' said William.

'Should've run for it,' said Jack.

'No, we were scared, to tell the truth . . . I won't ever go back.'

'Better not, it's not respectful. None of us would ever do that. It would have been a burial site,' said Arana. Her eyes burnt into his with hostility.

Johnson felt terrible; it was stupid to tell that story, no respect. He nodded and remained quiet until they reached the vehicle. He cursed himself.

By noon it was hot, and the family wanted to go home. They walked back to the car quietly, but Johnson's eyes were on Arana's backside swaying before him, and it was so round and beautiful he nearly wept. He imagined her naked and leaning over the bed and smiling back at him with an open mouth, panting for him.

The family stood outside the car and turned on the air conditioning. They opened the doors to let out the heat, then squeezed into the vehicle and drove off.

Wayne watched them go in a puff of dust. He kicked at a stump and shook his head. 'Idiot, idiot gunjie, that's what I am.'

CHAPTER 19

26th May, National Sorry Day

It was National Sorry Day and they went to the Wilga Land Council for a ceremony and a good cry as always. Aunty Mary sobbed through the reconciliation speeches and June held her hand. She would never forgive what happened to them all. Jack sat miserably on the side and said nothing. Kinchela Boys Home and starvation, beatings and bringing in cows with his feet frozen. It was too terrible to remember.

Later, at home, eating some tomato soup, everyone was subdued. The family sat in the lounge room and discussed the event.

'They are watching us, I'm tellin' you,' said Jack. 'Spies on the internet, and phone, reading all the stuff Rara puts up. Get her family into trouble.'

June watched Arana getting angry again, and worried that the girl was too passionate and might get involved in some silly action.

'Some high school kid's group is getting strikes happening every Friday for climate change actions,' said Arana.

'Don't get any ideas,' said Merle.

'Saving the earth comes first,' said Arana.

'Listen!' barked June. 'Don't ever forget that one person has died in suspicious circumstances already. I am investigating.'

'We don't know he's dead!' said Arana.

'He's dead, believe me! This is a dangerous activity. Big money, powerful people, corrupt politicians: don't risk your life.'

'We have thousands of followers on Facebook.'

'They watchin' us, all those bloody speeches. You women should stay home. I'm tellin' you,' said Jack as he paced the room flicking cigarette ash.

'Calm down, old fella,' June said, patting Jack on the shoulder, and she hummed a tune. She gestured to Arana to stop stirring Jack up.

'Okay, sorry. I'm worried about Steve speaking out,' said Arana.

'He did well, we like him!' said Merle. 'But, no, he can't stay the night with you. Over my dead body!'

'He loves me.'

'He's a man,' sighed June. 'The irrigators and gas consortiums might just murder him like the other one.'

Arana gasped, and a sob broke from her chest, and June hushed her.

'Cut it out, sis, you're scaring her,' said William. 'Be sensible, but the conservatives are at the meetings, the government spies. Even a CIA Yank up the back, trying to pretend he was from the Mormons. He had a book and everything, what's it called? *The Book of Mormon*. Narr. We should pick up old man Wright and that bikie Bam Bam and make them dig for his body.'

'Shit, that's daring,' said June. 'I'm all for it. Soon. But now, I'm going out for a while.'

'It's late and cold,' said Merle.

June opened the door to her car and felt the sensation of being watched. She started the engine and smoke rushed from the bonnet, and a stink of burning oil filled the air. She thought she might have forgotten to put water in the radiator and had busted the head or something. Flames shot out of the engine, and she dived out of the car. Cascades of fiery lights burst around her.

'Shit, help!' June screamed.

William rushed out of the house with a fire extinguisher and put the flames out. He stood back and with one arm pulled June upright from the road.

'You put oil in it?'

'It's been serviced.' She shuddered because it looked like someone had booby trapped her car. Horrifying; her head and body and heart shook. There in the engine the battery had a loose positive.

'Yep, thought so,' said William. 'The terminus has been exposed to something flammable like oil. A detonator might have started the fire. Lucky there's no fracture stick, you'd be cooked.

'Someone wanted to kill me?'

'Nah, frighten you.' William strode about growling, 'I'll get those bastards! I'll smash them to hell.'

June rang the police, but was not surprised when they did not show up. Things like that happened every pension night in Wilga.

* * *

After a few hours, June calmed down and borrowed William's car to drive down the dark track to see old Mary. White-painted stones showed the way through the Reserve. However, the neighbours said the Elder was staying in town and wouldn't be back for a few days. June thought it would give her a chance to tidy up the shack while the old lady was away.

Once inside, she jumped in fear at the sight of a large man in the dark kitchen area with a kerosene lantern. The room had a wooden table with tin cupboards, not even a sink. He was washing up. He leant over the plastic bowl with a mass of froth up his arms. He was scraping a pot and didn't hear her enter. It was Bundi. She shone a torch on him. His muscular arms were covered in froth. She breathed out.

'I didn't expect to find you washing up,' said June as she hesitated at the door. Her face pulsed with confusion.

He turned and looked her up and down. His eyes looked through her cotton dress and stopped at her throat. She pulled her cardigan over the cleavage. He inclined his head to invite her inside.

'Sorry to give you a fright. So, Detective June is on the missing person case, eh?'

'My clients tend to be angry wives hunting shithead husbands.'

'I help out here with Granny when I can. I'm not a bad man.'

He stood with arms dripping, and bubbles floated into the air and over her face. A tuft of dark hair stood out on his chest above the pale blue shirt. He leant towards her and touched the froth on her cheek.

June froze – she was close to bolting. She heard the clatter of bingo balls as the numbers were called by women under a gum tree. They loved a jackpot. They were so close. She didn't know what to do. She was mesmerised by his eyes. Her gaze followed his finger as it traced foam slowly down her neck to her shoulder. She felt a rush of heat. A throbbing in her thighs. There was no electricity in the shack, just lamps and candles, but it was comfy with plastic flowers and neat chair covers. Mary had refused to be moved to one of the new houses on the Reserve.

'You want to help dry? Take a tea towel.' He laughed and handed one to her, and she obliged by wiping dishes and putting them into the suitcase that stood for a cupboard. Outside, children ran past in the dust with tin-can trucks. It was dark in the corrugated iron room. A lace curtain blew in a breeze. A yellow kerosene light made the night magical.

He moved closer and leant his head on her shoulder, and breathed her in. He moved back and finished the dishes, sighed and placed the last dish on the bench. A canary in a cage slept.

'I have always wanted you,' he whispered. He moved close to her. His grey hair smelt of shampoo. His palms were open in supplication.

'Don't . . .' she said.

'I'm totally reformed. I abhor violence against anyone. Believe me, I had to fight to stay alive in gaol, but now, I want to love you. You wrote to me in gaol, have pity.'

June's heart raced. Her hands clung to her skirt, but he took the cardigan off and unzipped her dress and slid it down to the floor. She gave way, powerless and pulsing with desire. He knelt in front of her and buried his face in her belly. He reached gently for her breasts. He cupped them with his hands. His face tilted back, and he looked into her brown eyes. He gently stood up and kissed her throat. He moaned. Despite everything, June could not resist, and she tousled his hair, pinching the tresses in her fingers. Her legs shook as he unzipped his jeans and peeled away her knickers and took her standing up, face to face, with his tongue in her mouth. A passionate embrace. They panted urgently against each other for a long session of love. A deep shuddering sigh. Her head rested on his chest. She moved away.

'This can never happen again,' she whispered.

They dressed and laughed. June was embarrassed and kissed his cheek and walked out of Mary's house. She still had the tea towel over her shoulder and smiled at the bingo and card players and sat in her car, praying Bundi would not tell anyone. She would be mortified.

She switched on the engine and air conditioner, smoothed her skirt and pulled her knickers straight. She reached for a can of Coke, but it was hot so she tossed it over the back seat. Looking into the mirror, she rubbed the smeared lipstick from her lips and poked her tongue out at her reflection. What on earth was she doing?

There was Bundi, strolling from the tin house, whistling. Her moot tingled. She ducked down and hunched in the sticky seat.

June waited until his flash four-wheel drive was out of sight, then drove away grinning.

Back home, Merle looked at her with curiosity. 'What happened?'

'Bundi,' said June.

'Nah, you're joking?' said Merle.

'Nothing,' said June and she leant against the sink and smiled.

'His other women, like Betty, will be after you with a stick.'

'He's an awful man. Terrible, the enemy,' sighed June.

'I know.'

'I'd rather turn lesbian. Soft lips.'

'You do that, girl,' said Merle. 'Good, was it?'

'Unbelievable,' said June, and she burst out laughing with her sister-in-law. 'He can really drive you.'

'Nahagar, you'll go gwarnghi.'

CHAPTER 20

31st May

He even had a cross on his face. Bald head and snake-like pale-grey eyes. He wore black leathers with no insignia, and he moved like a bull. The bikies became quiet, even dropped their heads to their chests. But Bam Bam stood up to shake the visitor's hand. Outside the pub, June sensed the dealer's power; he was very dangerous, and suddenly he stared at her and her skin crawled. Then he disappeared. She turned back and he had gone. Just like a puff of smoke. June couldn't imagine how she would ask him any questions.

She imagined the scene: 'Oh, Mr Ice Man, killed any young blokes near a sacred site lately?'

Perhaps the question was best asked of the lesser bikies, who might like to gossip when drunk? Like Cowboy.

* * *

Next day, June woke at dawn with a sudden realisation that Bam Bam and Warren Wright might be working together. Even more strange was the fact that Bundi was a grandson of Mary's. Why was Wright so friendly with old Mary? Wright once bought old Mary a motorised mobility scooter, but she left it out in the rain.

If only June could force these suspects to confess. She had wild thoughts that kept her awake, a surge of woman power making her invincible. She'd gathered evidence from various interviews and now it was time to bring it all together. A certain letter from an anonymous person, spelt out in cut-out letters from a magazine, that the main drug dealers in town were implicated in Thommo's death. Now she had to be sure she had all the names. There were only two.

Suddenly it all seemed clear. The bikie was not the brains behind the murder and cover-up – there must be much more powerful forces, with sinister methods, who did whatever they wanted to get their hands on the millions of dollars involved in water business. Bam Bam was just a paid thug. She might get some truth if she had them together to talk. How could she force them?

A few days later came the miracle breakthrough in the mystery. A white bone stuck upright and pointed to the sky. Some Murri children on dirt bikes had seen this one finger poking up through the dirt, and had told June. But animals had been scavenging and scattering remains. The children told her, and she let William know. He had the idea to bring young Steve along to investigate without the cops. June said no, he could not come. It was against the law. It was reckless.

June took her phone and called Wright and Bam Bam. She told them she had something to say about water money and they should meet her at a certain place. And, strangely, they agreed.

* * *

Some hours later, June and William carried out the crazy idea of digging up that body in the bush. The sun was high over the treetops, the mud-caked shovel clanging in Mr Wright's hands. Then he stopped and stood miserably by the grave. He sniffed. 'You'll regret this, it's kidnapping.'

'You're interfering in police business,' Bam Bam hissed. 'That's a crime.'

June knew the bush places to search. She watched Bam Bam carefully, but he sounded amused rather than concerned. He sat on a log and smoked yarndi and giggled to himself. Almost blasé. They were way out in the bush, a fair way from Wilga, on a disused sheep paddock. The landscape was littered with piles of dung, and fragments of merino wool blew in the breeze. June picked at the fluff. Steve leant against a rock with William beside him. They looked triumphant as William held a rifle pointed at the men.

'Keep digging, Mr Wright,' said William. 'You can't just pretend you had nothing to do with this murder. He's there somewhere; dig him up so the goonge can find you.'

'What is your worst mistake, fella? What wrong have you committed, eh? In every darn lie there's a bit of truth,' said Wright.

June laughed. 'Think of it as moral choice here, a chance to redeem yourself.'

Wright stared at June, his eyes like coals. She had a hunch he was responsible for ordering Thommo's murder. And he knew that she suspected him, knew she had some evidence, and so he was dangerous. Was he mixed up in drug dealing as well as water thieving?

The pebbles showered down like confetti at a wedding. June felt something go hot in her head, like a coal of dread. Bam Bam's eyes drilled her. She noticed for the first time he had a tattoo that said 'I will suck your fucking blood' on his arm, and tattoos of skulls on each knuckle.

'My brother Gobblers will be looking for me about now. You will be shit when they track us down. They'll shoot and bury you both on top of old mate here, nice and quiet out here, nothing but galahs. I didn't kill the shithead! You reckon that little pea shooter bothers me?' Bam Bam shovelled.

'Let me try to understand this – you claim you're not involved?' June sweated.

'I know you find it a challenge, but some things need to be spelt out carefully to you Aboriginal inhabitants. I realise you don't have the education we have,' said Wright.

'Shut up,' said June. She looked exasperated.

William nodded at the dirt and Wright stuck the spade in.

'Let me explain something most respectfully to you.' June said. 'If we don't stop the irrigators pumping all the water from our rivers, as humans we will all die. The entire planet will blow away like dust.' She drew in a breath and her hands pressed against her hips.

'It's nothing, a bit of irrigation – it's called farming. We have rights, too! You exaggerate,' said Wright.

'You calling me a liar?'

'Just colourful hyperbole. Want me to spell that for you?' said Wright.

The two men kept digging, and at last they could see the bones with bits of dried flesh sticking to them. Torn clothing hung in shreds and from the skeleton, as if something had been eating it. The smell was ugly, like a stinking horse carcass. Flies flew around in a cloud.

Bam Bam threw the shovel into the grave and huddled on the side. William held the rifle up to his shoulder and aimed it at him.

'Better be ready to use it,' Bam Bam said. 'Because if you miss, I'll beat the shit out of you – and screw your sister.' He grunted and spat a green gob onto William's shoe.

'We should call the cops,' William moaned.

'That so? I know most of them, all me old mates, I'll get off,' snorted Bam Bam. 'You know. We're blokes, and we like getting our way. I kidnapped my first ex-wife and set fire with a can of petrol to her fucking house up in Tibooburra after bashing her new limp-wristed partner. That was fun. Whole fucking farm went up. Talk about global warming. He drove twenty kilometres in his little tractor to get help. Putt putt . . . Pissed myself laughing.'

Wright stood nearby as the skeleton stank in the sun. He was miserable and, terrified, he pointed at the bikie. 'I should never have listened to you, I thought you were smarter than this, but you're just a stupid thug.'

Bam Bam shook his fist at Wright and growled, but Wright looked beaten. He didn't want to believe he had been implicated.

William indicated to Bam Bam that he should keep working.

Bam Bam stared. 'You could shoot me now, and I wouldn't give a shit.'

William drove Bam Bam home, and Mr Wright shook in the back seat.

'It's a crime scene,' William said. 'We'll let the gunjie know about this.' He looked exhausted. The desert night air was freezing.

CHAPTER 21

1st June

Early next morning, Arana blew frosty breath on her hands and walked from the car to the stony sheep property near the sacred site. June and Arana leant towards the grave. Poor sweet Thommo. All hope was now gone. June sobbed and Arana held her back. The grave was surrounded with blue tarp and red danger tape.

Red dirt clung to the decayed body. Sergeant Blackett stood nearby and took notes on a clipboard. June's young cousin from Dubbo was now a forensic science graduate, a police officer, and she was the expert brought in by the senior sergeant and the higher police authorities. She had just completed her minimal general and field placement duties at a metropolitan police station. Everyone wore hazmat suits and masks and gloves. The Forensic Evidence and Technical Service had flown in that morning, and they bristled with importance. Providing high quality forensic services to assist investigations, prevent and disrupt crime, and serve the justice system: the website said so.

It was June's good luck to be allowed to observe near the site of the investigations. Strictly not allowed, but in the outback it seemed to be of no concern if an Aboriginal Elder wanted to watch and protect a sacred site. The Aboriginal Land Council was sending a site officer, but they had been delayed, so Arana wrote a report on her iPad, and was respectfully silent.

June was full of confidence; the body had been found, and she'd helped uncover the crime. Her cousin might be avenged. She watched the proceedings intently; all fear of reprisals had vanished. As though Thommo's spirit had called to her and filled her with power.

The issue of Mr Wright or Bam Bam laying charges against June and the family seemed to be resolved. They laughed it off because it suited them to avoid any difficult questions from police. No further action. In the case of interfering with a crime scene: they would have to wait to see if anyone would be charged. But Blackett shrugged; he didn't care anymore.

The forensic officer held the bones in a clear plastic bag. She laid the bag down on the ground and prodded it with a pencil. She was looking for carpal bones among the pebbles, and she meticulously separated them with tweezers.

'Looks like some bones are missing: the proximal phalanges,' said the officer. 'We will use biometric means, DNA and other means to assist in this criminal and coronial investigation. Blunt force trauma, by the look of it – a single blow by sharp implement, a tool maybe. Fractured left temporal parietal skull, epidural haemorrhage, probably. Great power in blow delivery.'

'Must've been a man who did it, then? Won't be any fingerprints in this mess.' Blackett turned over rocks to see parts of the decomposed body. Spiders and big black beetles scattered over dry leaves and he shuddered. Even June pitied him. 'The killer must have brought a weapon with him,' he said.

'Some bones are over there, I reckon, be respectful.' June pointed to where an animal had been digging.

On his first investigation of a real crime, the young, gormless Constable Johnson had an alert face and concerned smile. June saw him looking at Arana. He accompanied Blackett while he carefully scooped dirt from around the grey bones. A tray nearby held the victim's personal items: a wallet, a pen, a cigarette lighter, a silver signet ring and elastic hair ties. The putrid t-shirt lay on the tray with shredded jeans. Several cigarette butts had been collected, and they sat in a plastic bag.

'Stand back, madam. I don't know why you're sticky-beaking around this stinking job. You're not really qualified, are you?' said Blackett.

June shrugged and held up her Aboriginal Land Council badge as if it meant something important. Her hair stood up with the presence of a spirit. She knew that his body was protected by law and belonged to his traditional family.

Blackett grunted. She handed him a sheet of paper about respect for Indigenous Human Remains and protocols. She also held paperbark to cover the remains.

The sergeant used a pair of barbeque tongs to spread out the sandshoes and fragments of socks. He wiped his face and then retied a disposable mask around his mouth and nose to cut down on the terrible stench of a decaying corpse. 'No other fatal injuries by the look of it, officer.'

The forensic expert laid out the bones along a cleared place and nodded her head. 'Just a smashed-in skull. There would have been a lot of blood due to the impact on the skin over the skull.'

'Shovel?'

'Looks like it,' said Blackett. 'Not that I'm paid to speculate.'

The forensic officer scribed her notes and sat down in the shade. 'Maybe the killer was digging up something and was disturbed by Thommo and he hit him to protect his own identity.'

'Manslaughter, then.' Blackett sighed and looked away. 'You reckon you saw it happen, June? Well, better watch out. He might come for you too, or your family.'

'The victim has been moved,' said the forensic officer. 'Predators have had a go at the body – it's been gnawed and chewed up. Maybe a dingo has dragged off a leg; can't see a whole skeleton. Greenish bones and clumps of hair in dirt.'

Johnson called out from a rock shelter nearby. 'I reckon I found the leg bones.'

They all walked across the stony ridge to the spot, and there were crunched-up fragments of bone all over the floor of a low-roofed cave.

'We have to make sure it's not an Aboriginal burial site – you know, protected by legislation,' said Blackett, but June shook her head and photographed the site with the iPad.

'Nah, it's too fresh, those bones aren't old enough,' said Johnson. 'But all respect, Aunty, the Aboriginal drawings in that nearby rock shelter tell us that it's an important place, eh?'

June nodded, and stood back to let the forensic officer spread out her blue plastic canvas, body bag and implements. Blackett directed Johnson to lay out the stretcher to carry the remains back to town.

'Why would they kill him?' June speculated about the death. The skull was a particularly gruesome discovery: the open mouth and gaping teeth with shattered bone. It was almost certainly to do with dealers, or revenge by someone for something. Arana said she had her suspicions, which rested on a number of possibilities about drug dealing, but she kept quiet as the police might have tried to implicate her if she told them too much. The rumours from the mission were rampant. They whispered about the Ice Man. She doubted that the police respected any of her ideas. But her cousin police officer was shown respect, and that was enough.

The body of Thommo was boxed to be taken to the city for degreasing and de-fleshing. The autopsy would be done in accordance with the strict protocols of New South Wales Police. Dental records would be found, but there was no doubt in June's mind that it was Thommo. However, Blackett said he was mystified about the lack of drug-related evidence; everyone said he was a small-time dealer as well as a passionate environmental activist.

'Nothing to use as evidence against anyone including the bike club gentlemen,' he said.

'Must be something,' said June.

'No identification as yet – the corpse could be anyone.'

June wrote notes in her book, and was grateful that Mr Wright had not had them arrested for kidnapping. He tended to avoid formal police work.

Blackett shook dirt from his uniform and shooed June and her family away from the scene.

The death smell now clung to June's hair, body and clothes, and Arana looked like she would throw up at any moment. Crows and carrion eaters were all over the red rocks, pecking at fragments, and Johnson shooed them away. His eyes flickered over Arana, and he offered her his clean, pressed handkerchief. She refused.

June and Arana left the scene to the experts and drove home in silence. Then June sobbed. 'My fault, asking him to help me. I'm responsible.'

CHAPTER 22

3rd June

Gloria Wright, President of the Wilga Chamber of Commerce, didn't remember it was Mabo Day, she just cared about her business and online dollars; she didn't know about the end of terra nullius. June called around to her house after the phone calls went unanswered. She spoke through the security door. 'Mrs Wright, please, can I ask you a few questions? We want to clarify the actions of people from our town who might have seen something the day Thommo was killed.'

The door flew open, and Gloria stood in her bodybuilding lycra, her muscles glistening with olive oil. 'I am not to be disturbed when I am on a webinar, teaching body work!'

'Sorry, we tried your landline many times, but perhaps I had the wrong number.'

'Come in, for heaven's sake, I don't want nosey neighbours looking.' Gloria ushered June inside.

A chandelier shimmered over the entrance to the vulgar but lavish velvet-draped sitting room.

'Tea? Vitamin supplements? Steroids? Joke, Joyce. What do you want?'

'Okay, I'll jump in. We are trying to investigate our cousin's actions before his accident. Did you or your husband see Thommo in the days before it happened?' June coughed at her own audacity.

'No, that lad was a nuisance. He committed criminal acts against property, but I didn't really know him. He was not in our social circle, so to speak. Now, will you leave our house before Warren comes home and gets the shits with me for letting you in.'

'His partner said she saw you dropping Thommo off in your Mercedes. The day before he died.'

'Yes. Hmm. That was a private matter. Don't mention it to my husband, please. You know, that lad worked for lots of women, if you know what I mean. He was very good-looking . . . I helped him with his bodybuilding regime to keep the ladies happy.'

This was a surprise to June, and her mouth dropped open. 'So he was with you?'

'Just for an hour. For heaven's sake, don't be a prude – we're sad he's gone, too. A real Don Juan gigolo. Everyone has to make a living, darling. He was quite sweet, actually. You are very naive if you think this doesn't go on in a small town.'

June was stunned, and soon found herself outside the house. She noted that the Mercedes was covered in thick dirt and red dust. She clicked a photo and drove away. She wondered if the vehicle had been along the red dirt track where Thommo was murdered.

* * *

'I can't get that dead cousin out of my mind. I'm going to ask a lot of questions at the Knockout.' She looked pointedly at Jack, who paced back and forth.

'Coming for a couple?' Jack said. 'Look around, eh, William?' He swayed by the door. He moved from one foot to the other.

William hung his head over his mug of coffee. 'Too early, brud.'

'All the Koories and Murries are in town this weekend. All the whites left. I should have gone with them.'

June laughed.

The women busied themselves, washed plates and clattered about. Jack wiped up some cutlery and put it away.

'Give me that tea towel, Jack,' said June.

'No, I'm a feminist.'

'A huge number of gunjies at the football,' said June. 'There's twenty reinforcements being brought in. They're going to help that old gin jockey, Sergeant Blackett. I reckon he's only got a three-inch booble.'

'How would you know?' said Merle.

William put his head down on the table.

'Coming down to the Railway pub?' said Jack.

'In a minute. William can shout me a nice lemonade. I will use my best investigating techniques to find out around the pub about that fella. Might find myself a man too.'

'You're not.'

'Well, I might be fifty up here but I'm fifteen down there,' said June.

Jack screwed up his face.

'Wait till you see the hearts in the ballroom,' she said. 'The sheilas getting around in Aboriginal designs.'

'Staying out, June?'

'I am thinking about it.'

'Come on, I'm ready for a charge, real balal,' said Jack.

The family all sighed in unison. No chance to even keep a bottle of sherry to flavour the custard trifle. Jack would sniff it out.

'Bundi's in town for the meeting. He might pop around,' said William.

'I'll lock the door,' said June.

'He's all right,' said William.

'He might have some ideas about this mystery,' she said.

'Don't go sniffing around him,' said William. 'He's in bed with the Murray–Darling Basin Authority. I'm not asking anyone anything – it's trouble. I'm mooted.'

'He's in bed with a lot of people, I hear,' said Merle.

'I reckon I can find out some clues if I ask certain people.'

Jack tapped his nose and waved as he walked through town in rising midday heat. He whistled without a care in the world.

***

William sat outside the house with Bundi, smoking on a milk crate. June shut the back door and latched it. She was full of confusing contradictions; she couldn't face Bundi. 'I don't care if you have helped organise the river meeting, you can big-note yourself on a milk crate!' said June through the open window.

'I can help put up more posters,' said Bundi. 'We have to change the apathy. Blacks fighting Blacks ... it's no good ... There's white people who are paranoid; they've got secret fantasies of destruction. Murder their own over nothing. There's a sign out of town with a target drawn that says "You're now entering coon country. Kill all coons",' he said.

'We live here, we know,' said June. She listened through the flyscreen door, drinking a can of lemonade.

'You been to see Aunty Mary lately?' Bundi asked.

William cracked a beer and passed it to Bundi.

'Mind if I stay for a feed?' said Bundi as he winked at her.

'I would mind,' said June. 'But you might tell us what you know about Thommo?'

'He ran into bad trouble, for God's sake. He fought with his missus; she did it. Probably wouldn't let him get a doori. Someone had it in for him.'

'You know nothing,' said June.

'She's tough,' said William.

'I'll get a pie,' said Bundi.

'You do that! And buy a white ribbon to protest violence against women!' yelled June as he left the yard. 'We see the future, and it's female.'

'I'll wear it on my lapel.' He smoothed his polo shirt and twisted the silver and onyx signet ring on his finger.

'No. Tie it around your little booble,' said June.

'You'd know,' said Bundi.

'You went to prison for violence.'

'Lot of innocent Murries in lock-up,' said William.

'Eh, I got that sweet letter from you when I was on my holiday inside. You didn't think me awful the other night.' Bundi grinned as June slammed the window shut.

Her skin stretched to hide a smile. She walked outside and sat next to William.

'I've got a bit of a dilemma. I got an offer for a lucrative job,' said Bundi.

'Well, take it. You can look after your family,' said William.

'Big responsibility. Can't talk about it yet, but I'm tempted.'

He looked sideways at June, and she had a sudden dryness in her throat, and the terrible realisation that she cared for Bundi.

'We trust you, cuz.' William smiled. 'Don't mind Junie.'

'She's a gorgeous woman,' said Bundi as he whistled, and June smiled back at him and wanted to reach out, but stayed still.

Bundi slunk away down the road while June swept the backyard. He actually made her feel beautiful.

She knew that to the men in the community, Bundi had a glow like a heavenly apparition. He knew old ceremony and men's business. A man of high degree. People were quiet when he spoke; it was inexplicable. A calm floated from him. He knew about ceremonial things that other people only guessed at. He probably knew exactly what happened to the young fella. Knew why he ended up murdered. She wondered if Bundi could be talked into telling her. After all, she did have an investigation business.

Bundi once had a reputation as a violent man. It was hard for Aboriginal women to acknowledge that their men behaved like that. He denied it was him.

June heard that the police got him on some charge about violence and he was represented by the Aboriginal Legal Service solicitors, who were good at arguing for him to an all-white jury. He only got three months. June couldn't believe it. They even caught him with a rifle. He was a slippery fellow, all right. He had a roving eye and the wolfish grin that seduced. June pushed aside thoughts of him – he was dangerous. But sexy.

CHAPTER 23

5th June

On a cold morning the town park was empty. Some tufts of dry grass bent in the wind as June walked across the parched ground, and it came to her that the murderer might want to kill her or the young activist Steve who was taking Arana out. *Nosey Blacks*.

A text appeared on her phone. 'Fuck you, black bitch detective, stay out of cop business! Better watch out, you might get shot.'

June shuddered and looked around. The anonymous message was one of many. She had saved them, but she could not work out who'd sent them. If she replied then they would have her. She didn't trust the gunjies to help her. She felt like a victim and couldn't bring herself to share the fear with her family. At any moment she might be bashed or killed too.

Bam Bam rode up on his prize Indian motorcycle and carefully parked it under a blooming jacaranda tree. He took out a handkerchief and wiped some dust from the black chrome. He put his helmet under his arm and walked up to the playground and stared across at June. She ducked behind the swings, but it was useless – he could see her; he had followed her. She stood still and thought she might lie down and pretend she was asleep under a rosebush. Pretend she was mad, or drunk. Her heart raced.

'Gidday, miss, mind if I sit down for a mo?' said the very polite bikie as he poked his head around the rosebush. They were bright red roses, lovingly cared for by the council for the war memorial garden.

'What do you want?' said June.

'Nice weather we're having, eh? Congratulations on Walgett winning the footy. Might I just have a word? It won't take a moment.' Bam Bam smiled his nicest smile as he hitched his leather pants up over his crotch and squatted down. He scratched his knuckles where scabs oozed.

'Um, okay, but I'm alone here, so don't get too comfortable,' said June.

He took a pack from his back and produced a thermos and two plastic cups with Goofy on them.

'How about a cup of spearmint tea? I'm parched. Do you want manuka honey?'

'What is it you want? I'm not having your tea. Might be spiked,' said June.

He sipped from the cup and wiped his moustache with his fingers. He sucked the ends. 'That young pretty girl, Arana, she up home?' he said.

'No, she's not, and what's it to you?'

His eyes moved around the park, checking for anyone walking nearby. 'Nice spot, bit dry, though. I got a vague memory of you back in Redfern, you were a real looker. It's like old times.'

'Piss off,' said June. She stared at his hands. He would probably remember that she had spat on him once when he was a copper. She couldn't forget his sleazy smiles when he arrested the Koori girls outside the Empress Hotel. How on earth did he become a bikie? Then she thought it wasn't such a big leap. They were all in bed together. He dealt meth and grass and did the occasional break and enter.

'You shouldn't interfere with this cotton business, it's nothing to do with your dancing corroborees or telling Dreamtime stories. Stick to what you're bloody good at. That business of making me dig a hole. You must be fucking nuts. You don't want me as an enemy.'

'That right?' said June.

Bam Bam looked at the blue sky and sighed. She watched him carefully twist the lid back on his thermos. He saw a skink peeping from behind a rock, and swiftly grabbed it by the tail and bashed it with his boot. He laughed. 'You know, I shot my own father? He was a dick.'

'Jesus … Very informative talking to you, I might have to report you for harassment. I have to go.'

'Shot the old man. In the back. Just like he used to turn me over and have a go. Died like a dog.'

She saw his madness, and she quivered with the sudden fear. The fire in his blue eyes and the thought that he might pick her up by the leg, light as a wren, and bash her against a rock as well. With the hot westerly wind blowing emu shit in her hair.

He wiped the lizard guts from his boot on his filthy pants. 'Been nice talking, but.' He paused. 'You care about the locals, don't you? Some cotton farmers are doing it tough.'

'That so?' said June.

'You won't believe this, but I got a son at a private boarding school. It's a good school, you know? I've got high expectations. His mate is the son of Mr Wright. I want him to be able to hold his head up high, have morals.'

'You visit him, do you? Go to the parent cocktail parties?'

'Nope, his mum does all that. Rowing mother, you know. All dressed up in Jimmy Choo shoes. I grew up having the shit beaten out of me.'

'Sorry. Domestic violence hurts us all.' June gazed at the empty space of dirt where the carnival had recently camped. Dagwood dog stall and laughing clowns next to a flapping tent set up by the Aboriginal Land Council. Information about water rights and the New South Wales State government actions on water buybacks. *A load of bullshit*, she thought.

Bam Bam hadn't finished: 'I got something to prove, now. Nothing stands in my way, not you, or shitty activists or your niece. I don't apologise to anyone, and I'm not a fucking murderer.' He dropped his head to his chest and let out a deep sigh, then stood up and took out his phone and clicked a photo of her.

'Don't do that!' she said. 'How about you apologise to the son of the mission fella you murdered? He's a kid and will grow up without a father. Try telling him about your high moral values.'

Bam Bam opened a packet of Twisties and ate some.

'We finished?' she said.

William drove up and strode across the grass, and June stood and flapped her hands at William. She was trying to take command of the situation, show her power, but her legs began to shake.

'What the hell?' William said. 'Who are you? You barge into our park; you got a nerve! You stay away from my sister, and my daughter and her boyfriend.'

'Or what?' said Bam Bam as he flicked his Twisties packet away.

'Or I'll come after you,' said William.

'Beat me up with your what, boomerang?' said Bam Bam, 'Your ooga booga tribe will go on the warpath? Or you just going to use your walking stick? Did you know that Constable Johnson had an affair with my Duckie when he first arrived in town? He didn't know she was married. He's angry with Thommo because he hit Duckie and there was an AVO. Shove that in your pipe and get off my back.'

'Liar!'

The bikie laughed, and June leapt forward as William lunged at the big man. Bam Bam ducked as William pulled back his fist, his eyes shot with rage. June stepped in front of Bam Bam to stop the fight just as William landed his fist in her face. She staggered and fell, saw sparks flying, and she moaned and touched her jaw. She spat; her teeth were loose. The air was silent except for the heavy, laboured breathing of William, who knelt beside her.

'Oh shit, sorry, sis.'

'Good one, you should go on that blooper show, ha,' said Bam Bam. 'Now you've done it, bloody punched your own sister. Better go home and get her a packet of frozen peas.'

June sat up and held her jaw, blood running from her teeth. She glared at the bikie. In the distance she could see some young mothers gesturing towards them. They waved.

William turned towards Bam Bam. 'Get away from us! I'm going to call up the fellas at the mission and they will tear you to pieces.'

'You got it wrong. I wouldn't hurt a dumb kid all riled up about saving the planet, even if he was sabotaging the water pumps. It's not worth ruining my reputation.'

William sneered. 'You're lying.'

'I don't tell lies. No point. I own every action. No regrets. God forgives, the Gobblers don't.'

June had a tissue pressed to her mouth and William patted her back.

'We don't believe you,' said William.

She wished they would both evaporate, take their stupid male games and go somewhere far from here.

Bam Bam nodded and walked across the grass.

'I'll call the cops,' yelled William.

'Yeah, you do that!' Bam Bam laughed. 'You know the Sydney cops came out to our hideout and got us for a shotgun and one-point-five million dollars' worth of cannabis. Hit up for participating in a fucking criminal group. I got off. How about that? As I said, I'm now respectable. I give teddy bears to sick kids from the back of my chopper. I don't kill hippies or take water from your sacred shit river!'

June shook her fist at him.

'Get, I said!' William yelled.

'This has been fucking lovely. I owe you some payback, Aunty. One day I will come to collect. It's been nice reconnecting and remembering old times. I'm going for one big payout, and I can retire, go somewhere like Alice Springs.'

'And do what?' she said. 'What else can you do if you're not stealing something, dealing heroin and ice or killing someone and making certain that everyone around you is miserable? What good are you?'

'Hey, it stimulates the economy and creates jobs! Come round the clubhouse out bush, any time, sweetie, but bring your niece with you – she can have a free ride on the bike … Free rides and gifts like cuddly toys from Uncle Bam Bam, then she can suck me off!' He undid his jacket and tapped on the pocket. A big grey pistol stuck out of the cloth.

'You gunna use that, big fella? Or is it a water pistol?' said William.

'It fires, all right. Just don't make me cross. See you round like a rissole.'

'You got a licence for that?' William ran after Bam Bam and grabbed the back of his leather jacket, the colours of the insignia shouting 'Gobblers'.

The bikie elbowed William away with a swift jerk to the ribs and he leapt on the bike, roaring with laughter. 'Suckie, suckie,' he yelled and started up the engine.

They watched him ride off with a thundering roar. The wind blew a willy-willy down the road, and June thought that a little spirit man was making his presence felt. She stood with shaking legs and pulled William back from the road before he killed someone.

***

Back at home, June was trying to talk to a journalist. It was weird: the landline cut out, and while June was just looking at the screen on the laptop it went black. She thought it was odd – it didn't even restart. June looked at William; they had been told about an encrypted app, but couldn't work out how to use it. So if the police didn't want this information to get out, they were watching through the computer camera. Police had been eavesdropping on June's room. Phone calls were intercepted, and someone was looking into her room via the camera.

'You're just paranoid,' said William.

June emailed the New South Wales police, but they just replied that she was probably hacked by a foreign money scammer. She hoped her business computer was not hacked too.

CHAPTER 24

3rd July

It was National Aboriginal and Islander Day Observance Committee week, and the Aboriginal flag was raised outside the council chambers for the only time in Wilga. The family stood in line to bring smoke from the smoking ceremony onto their bodies, cleansing and healing the spirit. The local Wilga dance group painted up in ochres to perform and everyone cheered and joined in. Jack swung his legs in a shake-a-leg and everyone clapped. It felt good to celebrate the culture. William pulled some children up to dance.

'No shame! If you're shamed of our traditional dance you will be letting your mob down. Stand up and be seen.' He grinned as the little ones leapt into a goanna dance.

Steve joined in and was a fine dancer, while Arana led little girls wearing emu-feather skirts in a circle.

There had been a time when the mighty Ngiyampaa and Barkindji nations had walked the land and cared for it – for seventy thousand years. They had been shot and starved off their Country, or driven away to far-off places where stories, language and clans were erased.

At sunset, Steve stood out on the red dust plain watching the eaglehawk soaring against clouds; he imagined thudding corroboree feet. The earth hummed alive with spirits. The giant red kangaroos had been hunted and made into dog food. He looked at the scats and wondered how many marsupials were left. He kicked at some white bones on the ground and knelt to examine them. Human, maybe? But the jawbone seemed like a roo, so he moved on; then the old settler shack came into view. The mission fellas had given instructions about how to find it. Worn-out caravans stood nearby with peculiar plastic pipes sticking out the windows. Probably meth labs.

Steve noted the location of the hut on his mobile and walked back to his motorbike, a Kawasaki he had just bought cheap in Sydney. A mode of transport, nothing else. Not expensive to run. He wanted to show Arana a good time, mostly walks and meals at her parents' place. He would make it up to her one day when he had a full-time job.

It was a warm evening. He pulled up outside June's. Arana jumped onto the back of the bike with a yelp of delight. They could ride down to the riverbank, get some privacy. He had some condoms in his pocket. He was in love with her and would protect her from any trouble. But first they had an investigation to carry out. He told Arana that he wanted to follow old man Wright and try to find out some information.

'What are we?' she said. 'Private detectives? Like Aunty. I loved *Inspector Gadget*.' She squeezed his waist and breathed in his masculine smell. They putted off into the growing dark to the back of the RSL club and noted the big new black SUV that belonged to Wright. Steve motioned to Arana to wait with him by a big coolabah tree, large enough to hide five people. They waited, and she tried to hold her giggle and he smiled as she kissed his ear. After ten minutes they watched Wright stroll down the club steps while he picked his teeth. Beside him was Bundi and Sergeant Blackett.

Steve clutched her hand as Wright's voice boomed with authority: 'Under new laws introduced by the New South Wales Government, irrigation pumps on rivers must be fitted with telemetry equipment that will transmit "real-time" data so that farmers and authorities have transparent and accurate records of what water was pumped and when. These new rules help to show the irrigators are doing the right thing.'

'Yeah, mate, but we're in drought,' said Sergeant Blackett, 'and it's not a good time to be spending a lot of money on high-tech equipment. You want to contribute some funds to the police benefit program instead.' They laughed and shook hands. Blackett disappeared into the darkness and Bundi went back into the club. Steve froze as Wright looked over at Arana and him, but the tree's shadow covered them. Wright took out a packet of cigarettes and lit one and flicked the match onto the bitumen.

'Littering,' whispered Arana.

Wright unlocked his car with a remote and slung himself into the cabin. He adjusted the rear-view mirror to smooth his hair and drove off.

Steve waited and then kicked the bike into life and took off after him. Arana clung on as they rode fast after the speeding black car. The vehicle drove out of town and over the wooden northern bridge, before suddenly it swerved down a dirt track, and they heard the bump as it hit a cattle grid. Steve followed but lost sight of Wright.

After twenty minutes, Steve pulled over and checked his mobile for directions. He looked behind them and whistled. 'His mates have a shack out here nearby. The mission fellas told me about it, but I can't work it out in the dark.'

'We might have been seen when he was leaving the RSL,' said Arana.

'Nah, I think we missed a turn-off back at the last cattle gate.'

She asked him how he knew about Wright. Surely the councillor was respectable, at least to white people. Surely he was honest.

Steve laughed.

'Why are we following him?' she said.

'To learn the truth. One fella goes missing, a river activist, tree hugger – well, it might happen again. The cops don't care, but we might find out and expose the lot of them.'

'We're the underdogs. No one much cares about Murries – too many old bodies buried on Country,' said Arana.

A headlight of a vehicle crisscrossed the trees: someone had a search light for hunting, and they were using it. Steve flicked off the bike lights and they dismounted and crouched down on the ground behind it. A double thorn pierced Arana's knee and she yelped. Steve's hand pressed against her mouth with his leather glove. She clenched her sweating hands; her breath was fast; now she was scared.

Suddenly the black vehicle tore towards them, bouncing on the rocky corrugations. Steve pushed the bike out of the roadway and dragged Arana after him under a hanging rock. They waited; his breath pounded in his ears. The car slowed down, and light hit the bike frame, lingering, almost caressing the seat and pillion, but then turned off. Nothing to see, but Arana and Steve were frozen, as if the beam of light was right on them. They might be Black, but their faces were pale with fright.

The vehicle roared away, back the way it had come. It looked lost, too.

As they calmed, Steve meditated on the dimensions of finding the shack, of what he might discover, of the likelihood that it would contain evidence. He stroked Arana's back, and she breathed out.

'We go home now?' she whispered.

'Haven't found out what we are looking for,' he said.

'What if he realises it's you?'

'We'll be right – he didn't see us.'

Steve's eyes rested on the tufts growing by the track, cotton plants peeking through the dirt like invaders from another planet. White puffballs gleaming against the scrappy saltbush and dry wildflowers.

'Cotton. It's taking over the bush,' said Arana.

'Come on, he's getting away.'

'Take me home.'

'Too late.'

He pulled the bike upright and they took off, following the fading rear lights of the black car, gaining on it.

A kilometre more of the rough road, then Steve stopped the bike, and holding Arana's hand he crept through the bush towards the 1880s slab hut with its burning light. A generator huffed. The house was a lonely figure in the sparse gum trees. Around a fire pit, lounge chairs sat with stuffing hanging out. Piles of beer bottles littered the ground.

They slinked their way until they were tucked right up beside the house. They saw two huge black motorbikes, bodywork shining like silver.

An Indian Scout and a Matchless X. Nice bikes, thought Steve.

Pots of lush green marijuana stood in neat lines next to the fence, with barbed wire over them and a watering system operating from the plastic water tank. But the most interesting objects were the boxes piled under a corrugated iron lean-to. Steve crept over and opened one, and saw what he'd seen on a couple of large government infrastructure construction sites before, when he'd needed cash-in-hand jobs: explosives. He clicked a photo and put his finger to his lips to hush Arana.

She pulled her hand from Steve's and looked like she might run, but he was amazed when she instead held up her fingers to hush him and signalled that she would go round the back of the hut while he took the side window. She looked confident, but he saw her chest heaving as though she could barely breathe.

Steve watched her creep up the path and hide behind an old dunny. She moved to the side of the building and knelt on the ground and waved to him. He thought that perhaps she might be able to run from the men if they turned on him. She was a fast runner, a real Cathy Freeman.

He stood up to the window and looked in. It was like a coven of witches – no pointy hats but several filthy-looking bikies in ragged sleeves and tattoos. A rat's tail hung out of the grey head of the biggest of the two. Steve slunk down and prayed that Arana would not be seen. He put his ear to the crack in the pit-sawn wall; he could hear the men clearly; it was like some sort of Aboriginal cop movie, *Goldstone* and he was Jay. He pressed his knees together, worried that urine might start to trickle down his leg if anyone looked towards the area where Arana hid. He bit his lips and concentrated on the conversation.

Wright was standing with his leg on a low chair. He had a smoke in his mouth and looked remarkably relaxed, as if he was with brothers. Steve thought, Yes, that's it, he's in disguise in the respectable costume of a blue-lapelled shirt and riding boots. It's all con. He was crooked; the act with the Elders was crap. He might be a relation of old Mary, but he sure didn't act like it. He seemed like any redneck racist shithead.

Steve listened.

'I'm just going to say this once, Bam Bam,' said Wright. 'You'll get paid as soon as I do. They're waiting for the big cheque from the Chinese. Any day now, I promise you!'

The big bikie grunted and spat on the floor. 'A deal is a deal, you cunt. You know the rules for us one-percenters. You want me to cut off your tiny dick and hang it outside your wife's door? Or will the brothers just go around and root her for you? Or do you like to watch?'

'Here now, we don't need that kind of talk – you know I'm good for the money. I'll put half in your account tomorrow. Do you do internet banking? What's your account and BSB?'

'No payment? It looks real bad. Wassa. But we'll see how you measure up in the next few days. Cash, you cunt! You can pay off your fucking fuzz yourself.'

Wright muttered under his breath.

'What's that? You calling me a cunt, cunt?' said Bam Bam.

'I never said to kill anyone! Whose idea was that? I said scare him. He was sabotaging the water pumps. Would've cost a fortune to fix. Silly bastard. Dumb kid. Had a wife and little one.'

'Bring out your fuckin' violin. I didn't do it!'

Outside the window Steve opened up the recorder app on his mobile. It seemed too far away to hear properly but, still, he started recording. He listened intently, then one bikie turned to look out the window and Steve ducked and broke out in sweat beneath the sill. He prayed he was invisible, as his uncles had taught him when hunting. He held his breath, then moved inch by inch down to the damp ground.

Inside the shack, Steve could hear Wright opening a bottle of wine and carefully pouring it into stemmed glasses that he had brought in his pocket. Clink clink. The men drank as if they were at a Hunter Valley wine tasting evening; they swilled the liquid around in their glasses and sniffing.

'Now, that's a nice shiraz, where you get that? I'd like a dozen on the house.' Bam Bam laughed.

'Hunter Wine club. It's exclusive; you have to be invited,' said Wright.

'Well, fuck me dead – you can get me an invite, or you'll be pissing blood instead of wine.'

Steve bumped into a kerosene tin; it clattered. The men whipped around and Bam Bam drew a Glock pistol from his pocket and cocked it up near his cheek.

He moved slowly to the door and kicked it open. 'Who the hell is there?' he shouted.

A sound like a dog barking repeated in the air. The bikie swerved around, and suddenly ducked. A large barking owl flew out of the tree nearby and swooped over his head.

'Farrrk!' he yelled and fired his gun into the air.

Arana was petrified, and stood stiffly against a tree. Steve spotted her outline in the dark and he tiptoed up and touched her hair. She jumped. His eyes bulged, and he motioned to her to be still. Her face was buried into his chest. Hand in hand, they moved like a featherfoot over the track, softly. They went as far as they could from the shack and hid behind a pile of hessian bags. They didn't want to risk starting up their bike. She prayed there would be no snakes, though being bitten by a brown snake seemed preferable to being shot by Bam Bam the bikie.

They would wait until the men either left or were asleep.

After half an hour the door of the building banged open and Wright strode out to his car. He clicked the remote and waved to a man standing by the shack.

'Any other jobs you want,' Bam Bam called, 'better bring crisp cash, you little prick!'

'I get you, no worries, cash is king,' said Wright, and he strolled up to the pot plants to pinch off a resin-soaked head.

'Don't touch the merchandise, Wright, you stupid prick! Fuckin' unbelievable!'

Wright jerked his hand away and hurried to his car. He drove away at speed.

A few minutes later the bikies climbed on their bikes and roared after him.

Then it was quiet, and the owl began hooting.

Arana sobbed in Steve's arms, and he threw off the smelly hessian bags and helped her stand up. He smiled in the darkness and brushed the leaves from her hair. He hoped his little bike had been hidden from view. 'I shouldn't have brought you here,' he said.

'At least we know now. We can tell the police.'

'No, we can't tell the local gunjie, too risky.'

'His family needs to know,' said Arana.

'Yes, but Bam Bam said he wasn't responsible.'

'That owl is a bad omen, death is coming.'

CHAPTER 25

11th July

It was the Wilga Rodeo Day, but June didn't go – she hated seeing animals beaten and kicked. It didn't seem fun, just bloody cruel. She remembered the carny people in full pulse, and dagwood dogs everywhere, and luminous red sauce running down children's arms. Faces sticky with fairy floss. Even the clown with a silly hat running away from a bull in pain was stupid. She thought she might just turn vegetarian.

The town also seemed full of drunks and disappointed betting men. Their faces looked like someone had hit them with a shovel. The show was set up with a jumping castle and a shooting gallery of rusty ducks. Nearby, the local paddock was full of abandoned show bags and kewpie dolls on sticks and wrecked cars covered in animal shit. It was a day when winter dust blew in people's faces, their eyes squinting under Akubras.

Inside June's heart and head was a pulsing, deep-thinking Black woman, a person who valued self-deprecation and a lack of meanness. She gave to whoever needed it and showed generosity in most things – but she couldn't tolerate someone who stole from other people, like the TAB. Above all, she hated those who stole from nature and ripped off the environment so they could buy a new shiny BMW. Her impulse was to scratch the duco with her keys. Yet the feeling that was foremost in her was compassion. If she spotted someone crying in the street she couldn't walk past. And she had brought home countless homeless people to stay awhile until she found a refuge to take them in. 'A real Mother Teresa' her brother called her. All living things deserved compassion and empathy. Except the river robbers. They could rot in stagnant ponds with dead fish stinking around them.

Then there was the never-ending mystery about her cousin Thommo. She saw his little son walking to preschool with Duckie. June waved and thought that she needed to do more to find the people responsible for his dad's death. Poor little boy.

She never wanted to stand out from the rest of the mob, to be a big-noter or loudmouth. There were plenty of fellas to take those roles, but she could never sit still and watch people abused or set upon by bullies; she couldn't tolerate a thug having a go at a little fella or kid in a playground or a disabled person. She had often helped her mate with cerebral palsy cross the road or have a go at roller skates, always with all the kids jeering at him. He dribbled and waved a favourite stick all the time, but she knew him to be quite smart and funny; he cracked jokes that as a kid would make her wet her pants. Her dad always said, 'Stick up for the underdog,' so she did. He also said, 'If you're the only person in the room who knows something is wrong and all are against you, hold your ground, stick to your beliefs and don't be scared.' Her father's presence loomed in her heart; he was always with her, even though now long dead. His hands as they held a bird's broken wing and mended it with thread and a matchstick, the big dark hands full of concrete dust and stone scars. He had worked building walls, and any job on a farm that would pay a wage barely enough to buy a bag of flour, tea, sugar and a scrag end of beef. They lived close and warm, with only curtains to mark off the bedrooms in their corrugated-iron house by the river.

June thought how time was important – to seize the day and identify criminal suspects. Months passed and people forgot details. It had become too difficult to find credible witnesses besides herself; June was left almost thinking she had suffered a massive hallucination. There had been no dead Thommo, no body lying in gore. No actual evidence. It had all gone cold.

She strolled along the street. A sound like growling lions pulsed in the morning air, and twenty bikies in black leather roared through the town. The bikes were shiny and menacing, and June watched them pull up one by one and back into their parking places outside the Railway Hotel. Arthur came out to greet them. Then he spotted her watching from the corner next to the courthouse, and he nodded and winked at her. Among the gang was Bam Bam, wearing his full colours, and he acted like they had elected him the president.

June sat down on the garden seat and watched with her scarf pulled close to her face to keep from being recognised. The person who lived inside her mind suffered countless doubts, premonitions, spirit visitations and yearnings, all that controlled how she behaved in the outer world. The Aunty lived in fear that one day Bam Bam would get his revenge against her for the many humiliations. She imagined a carved target on her back etched with a sharp penknife. After all, she had kicked him in the groin; he would probably bide his time but come after her.

Cowboy climbed from his chopper and shook hands with several out-of-town bike club members. He had recently been patched, and he had the full set of insignia emblazoned on his vest. June wondered if Aunty Mary had done the embroidery. June knew that Bam Bam was the elected Sergeant at Arms of the Gobblers club; he was the Enforcer. But she suspected that there were few local members, and that the men in front of her were from Dubbo or Bathurst.

It was a hot-thirst day at midday, so June had walked into the pub and ordered a lemon squash and took it to the beer garden. She waved at the drinking women from the screen-printing art shop and sat down under a tree. They carried plates laden with pies and chips and leant towards the tables with hungry appetites. No one looked at the bikies; they were too dangerous. All the local drinkers pretended they couldn't see them. Except June. She twitched.

She saw the ice dealer enter the beer garden. She opened the recorder app on her phone. He was huge, with tattoos down his arms and neck. This part of the pub was a free zone where Aboriginal and whitefellas mingled. This was where she could listen in to conversations. Maybe someone would drop a few clues. She clutched the lucky silcrete stone in her pocket. It was full of ancient power; it throbbed.

This man had a cross on his face. What was he, some sort of Christian fanatic? Bald shining head and snake-like pale grey eyes. He wore black leathers with no insignia. He moved like a bull. The bikies became quiet. They dropped their heads to their chests. Bam Bam stood up to shake his hand. She could sense the dealer's power; he was dangerous. Then he stared at her, and tilted his chin towards her phone on the beer mat. It disappeared into her pocket as she turned around, hoping he was looking at someone else. Anyone else. She sweated and her skin crawled while she tied up her long, gold hair in a band. Then she fiddled with her Aboriginal flag earrings and pulled her skirt to cover her skinny legs. Her coffee skin and beautiful face shining. Just an old gin having a Fanta. A private investigator minding her own business.

He disappeared. She turned back and he had gone. Just like a puff of smoke. A shapeshifter, like a koradji. June couldn't imagine how she would have asked that bikie any questions about the murder. His eyes could melt steel. She had pictured herself with her bosom uplifted, a strong black woman.

The Gobblers were in full-throated conversation and the beer flowed. Bam Bam stood with one foot on the outdoor bench and surveyed his club members, his plaited beard tied with a red scrunchie. He scratched his balls and sniffed, then leant forward and welcomed the members to the formal meeting.

June was fascinated, and wished she could move closer to better hear what they said.

Bam Bam coughed to clear his throat, and began. 'We have had an interesting year. What with the killings of, um, chickens? You can taste the gore, congealing on cracked bone. I'm not one to brag' – laughter from the gang – 'but after I had a poisonous childhood, some of you remember my old man, oh yeah, he could smash a face, all right, like you one-percenters.'

The rest of the conversation was wacko military lingo that seemed to describe a shootout with unknown bikie enemies. Bam Bam cleaned his fingernails with a large pocketknife, clicked the blade back and put it in his leather belt pouch.

He spoke in a hushed voice. 'You know I'm a people person. I can't walk on water, but I can fuckin' shoot. I had a 7.2 Seiko with a fuckin' telescopic lens. Who fired the shot? Don't know. We did okay, eh?'

The ice dealer returned, and dropped a small bag of white powder on the table. June turned around to see it disappear into Cowboy's jacket. The dealer whispered in Bam Bam's ear and he looked at June. She hunched her body and wished she could disappear. Her hands trembled so she sat on them. The Ice Man looked at her and waved. Perspiration dripped from her face as her heart bashed her ribs.

Cowboy sat down and stared into his beer. The group watched the Ice Man coolly walk out of the beer garden. Bam Bam's eyes fell upon June. She doubted that the gang would pay attention to an old lady in a scarf and sunglasses, but she was wrong. A nuclear generator of fear rose in her body and lodged in her throat. Her lips stuck to her teeth; he'd caught her in an impossible moment of wanting to stand and run or hold her ground and pray for deliverance. Stupid. Stupid, she thought.

'Look who's still here!' he hissed. 'The black queen of snooping bitches. You've got a hide!'

June ducked down and crept along the wooden pub bench, but he was on her in a second and his foot crushed her hand. He released her. All the gang surrounded her.

Bam Bam reached for her glass and squeezed a large blob of tomato sauce into it and followed with pepper and a long dribble of spit. He pushed the glass under her nose. 'Have a sip, go on.' He smiled through broken teeth.

June felt hot. She fanned herself with a napkin. 'Not thirsty, thanks,' she lied.

Cowboy leant against her shoulder and sniggered. 'Come on, Aunty, we know you want to.'

She sat upright and sighed, her fingers grasped the glass, and the men breathed out. She tipped it on the grass and stood up. 'I have to go now – it smells out here, like something died.'

June shook, but her determined chin stuck out. She no longer cared if they hit her. She had been walking along the track of avoiding whitefellas like them all her life. They had surprised her with their reckless stupidity. She didn't want to ask herself why at this moment she could be brave, why life had taken this turn. It was a burst of awareness, some clarity among the rubbish. She breathed in the animal perfume: a mixture of stud cologne and male sweat. They hunched their shoulders in a wall of black leather. She had an image of climbing out of the pack like a sheep dog over a flock.

'Not even scared,' said Bam Bam. 'I admire that in a person who claims to be some sort of investigator. Respect it. Look, her eyes don't even blink when I look at her. Fuckin' amazing.' He paused. 'She's harmless and old, let her go.'

The gang moved aside, and June climbed from her seat and walked with dignity to the exit. Laughter poured behind her and perspiration stuck her skirt to her legs. Her hand clutched her phone, and she stopped.

Then Arthur was beside her, anxious and red-faced. 'You okay?'

'Yep, no problem.' Her face burnt.

He attempted to stroke her arm, but she flinched, as if hit by a cigarette ember.

'Leave it! You got CCTV in the garden?' June whispered.

'Yep, it's running.'

'Might be useful one day. Evidence.'

'I don't want any part of your investigation. The cops said it's over. They couldn't find an organiser of the crime, or executioner. These bikies deal steroids and stuff, no actual harm. It's a club; they enjoy male company.'

'They need to be reported.'

'They drink here, so I'm implicated. I hear things.' Arthur wiped the handrail with a cloth and moved back to the bar.

She went along Wilga Road, a quiet space with red earth mounds. The land was flat, and she thought it was a place that could hide a crime. Stones piled up in small cairns, and each one looked like Thommo's grave.

Behind her, the sound of a Harley-Davidson. She was utterly vulnerable on the outskirts of town. Stones skidded beside her as the bike tore past, then it turned around. The bikie was not familiar, an out-of-towner with a black Nazi-like helmet. He rode back and around her again and again. She wanted to run but it seemed futile. Heart racing like a torpedo. Bluish tattoos stood out on his face. His eyes were covered by black goggles and he had a military-style bayonet strapped to his leg.

'I'll be watching you,' he cried. 'You take care, Aunty! Fucking never sleep!' And off he rode with the dust billowing over her body. She spat the dirt out and wiped her eyes; saw herself run down like a rag doll.

That evening, she kept seeing Thommo's skeleton laid out on a gurney. Bits of grey-green bone sticking out. The stink, the vomit, and the nausea rising. She would run away to some exhilarating destination like Wollongong. Start a new investigative agency. Maybe work with Steve. Get a higher profile.

At home, she climbed into bed, her laptop in front of her to check if she had any new clients. Nothing. The pillows looked oddly piled on the end of the doona cover. Her favourite Koori print of a painting, a snake, Madukka, was turned to face the wall. It felt weird. Had someone been in her room? She called out to Merle and asked about it, but the reply came in the negative. June stared out the open window, and only then noticed the flyscreen slashed at the bottom with a knife or sharp something.

When she eventually did fall asleep, the once comforting Wilga home felt full of dangers. Was there someone outside her window now – a scratching noise against corrugated iron? Her heart beat like a cricket. She dozed, then sat upright. Mosquitoes buzzed.

June pushed aside the night fear and reached for her echidna beads and whispered a prayer, though to who or what she couldn't say. How could she go back to sleep? The cold sheet stuck to her back, so she got up and had a shower.

Her head was bursting with ideas about how to trap the murderer, but they all seemed ridiculous.

CHAPTER 26

28th July

A cold day like any other for Blackett at the Wilga police station, except for a morning visit by 'Detective' Aunty June.

He saw her walk up the steps, and he was about to exit out the back to avoid her when Johnson called to him. He listened with an expression of boredom on his face, but June was dangerous if she wanted to be; next thing he would be answering questions at the Police Integrity Commission.

She sat down and told him that worry about Thommo's death was upsetting the whole community. People demanded answers. Her client was now Duckie the widow, and June wanted him to bring the killer to justice. June admitted she had encouraged the deceased to meet someone and discuss environmental activism, nothing else. But she had a piece of incriminating evidence: Steve had given her the phone audio recording of Bam Bam talking about Thommo.

'Bam Bam actually admitted to hitting the young man with a garden tool,' she said. 'Listen to it. It's a bit scratchy. Take the USB stick. It's evidence. I have a copy in case you lose it.'

'You telling me how to do my job?'

June nudged the stick towards him.

Blackett pushed it to one side with a pen. 'It's safe with me, ma'am. One thing, though: falsely accusing a man can be seen as attempting to pervert the course of justice. You would have studied that in your course. It's an offence.'

Johnson stood up. 'You can be charged, but Sarge won't do that, eh, sir?' Johnson smiled.

Blackett placed the USB in an envelope and sealed and dropped it under the desk into a rubbish bin.

June looked with determination at Blackett. 'I need to know if the recording is clear enough. Might be faint. Surely it's permissible evidence.'

'No permission had been given to record someone in secret,' he said. 'It's probably worth nothing, could be wasting our time, but let's not keep you. Must have plenty of clients waiting for you. All with bulging wallets.' He opened the door to the office and escorted her towards the exit.

'Why did Thommo text you? Explain that. Please! Are you in with the ice dealer?' June almost spat at him.

* * *

A few hours later, Blackett drove to Bam Bam's caravan and found Cowboy there too. He took hold of some footwear and brought the bikies in for questioning.

The police interview was taking place in the Wilga police station. Blackett watched Constable Johnson deliver some takeaway coffees for the suspects. He was a kind man. He looked pathetically excited to be part of his first criminal interview.

Blackett indicated to the constable to set up the video camera on a tripod and press go.

'You took your time,' he said.

'I had to feed the emu. I call him Ozzie.' Johnson placed the coffees in front of the suspects.

'I'll shoot the bloody bird.'

Blackett sat down and nodded at Bam Bam and Cowboy, who sat with relaxed grins in the interview room. They wore blue workman's singlets and dirty jeans.

'Read that out to them,' said Blackett. He handed a booklet to his sidekick.

'You have the right to remain silent,' Johnson read. 'But any answers you give will be taken down and used in evidence against you in a court of law.'

'We demand legal aid. Like the Blacks get for free,' said Cowboy.

Johnson continued. 'These are not summary offences, sir, not just a minor traffic violation, but real charges. They are indictable offences, and will be dealt with properly in accordance with New South Wales laws.'

Blackett was not happy about this interview, but he had to follow instructions called in by city detectives. He really wanted to stay home and watch his wife vacuum the lounge room. He opened a notebook. 'Gidday, Bam Bam, Cowboy, nice to see you both again so soon.'

'I need independent legal advice before I answer any questions.' Bam Bam put his feet up on the desk and Blackett leant forward and looked at the sole. The bikie put his feet back on the floor.

'New boots?' said Blackett. He yawned and scratched his nose. He had a pimple coming.

'What about those old boots?'

'They had melted soles,' Bam Bam paused. 'Bushfire, mate, I was a volunteer, I hold a hose.' He laughed while Cowboy squealed with mirth.

'Okay, forget about your footwear. This inquiry may be terrorist-related. We have been given information by a certain private investigation service.'

'What would a gin know?' said Cowboy.

Johnson leant towards Blackett and whispered something, who nodded, and Johnson read from a folder. '*The Australian Security Intelligence Organisation Legislation Amendment (Terrorism) Act 2003* has made it an offence to not answer questions by investigators. New South Wales "evidence of silence" law allows courts to draw an adverse inference from silence.'

'Silence is golden,' said Cowboy.

'Shut the fuck up,' Bam Bam hissed at his mate.

'Terrorist activities have been identified, with River Action groups planning to blow up irrigation systems.'

Blackett nodded at the suspects and drew a picture of a car he wanted to buy in his notebook.

'How long have I known you, Sarge?' Bam Bam said. 'What, ten years? You know me. I'm a bike enthusiast and I raise money for the kids' hospital.'

'Sorry, Bam Bam, but we will ask you both some questions. The local central command has ordered it.' Blackett imagined a grog night by the fire at the bikie hideout. He also remembered Bam Bam when he was a probationary constable in Redfern back in the day. Not exactly mates, but associates.

'We didn't follow Thommo on our bikes. We didn't even know the fella. Ask his dealer, that Ice Man. Don't act like you don't know him real well.' Cowboy beamed. 'Anyway, you can't take any forensic evidence unless you arrest us. No fingerprints, no DNA swab, hairs, blood, nothing. Are we under arrest?'

'We already have all that from the last time you gave us a little visit.' Blackett grinned.

'You haven't got mine. Not giving it, anyway,' Cowboy shouted and rocked back and forth on his chair. 'We didn't do that, neither. That person of interest in that town cut his own throat with a razor. It had nothing to do with Gobblers. We got off.' Cowboy smiled and yanked his thumb at his mate.

'Can I ask something, sir?' said Johnson.

Blackett nodded.

'We need to find out if you fellas had the motivation and the method?'

'Everyone can get a shovel,' Bam Bam said. 'And as for motivation . . . Well, he was an annoying little prick with an earring. Gave every bloke in town the shits. Women couldn't get enough. I had nothing to do with him going missing.'

'Lying to an officer could result in an obstruction of justice charges. Serious time in prison.'

'I love it: catching up with old mates, free food, free blow jobs. All good.'

Johnson sighed and put a file in front of his boss.

'You should destroy that forensic stuff,' Bam Bam said. 'You could wrongly implicate me in relation to other offences. You could fit me up with rape or anything.'

'Did you attempt to rape Duckie?' Blackett stared at him.

'Did you? Nah. For God's sake, she's my ex-missus, she's got my kid. I love her.'

Cowboy guffawed. 'He's kidding. Anyway, she's a slag, town bike, everyone gets a bit for a deal of dope.'

'Shut up!' Bam Bam yelled. 'You arsehole, I'll deck you later.' He shoved Cowboy, who toppled from his seat.

Johnson helped him up and brushed him down.

'Don't touch me, Constable, that's assault!' said Cowboy.

Blackett looked bored. 'And there's the matter of a dumped vehicle, belonging to Thommo Smith.' He scratched his hands and slammed shut his notebook.

'Don't know nothing.'

'Fuck youse both,' Blackett growled. 'I've had enough of this shit. Tell us what you bloody know or I'll throw the book at you.'

'What do you care about another dead blackfella? Who cares?'

Bam Bam stood up, and Johnson opened the door for him to leave, but then the bikie sat back in his chair and hummed.

'Didn't do it,' he said. 'Don't know about it; innocent as a newborn; ask the Ice Man who done it. You're mates with him, aren't you?' He laughed.

Blackett sighed. He didn't want to do this anymore. He felt exhausted.

'Get me a cuppa while you're out there. Two sugars 'cause I'm so sweet.' Bam Bam crossed his arms and slipped down the seat. 'And a stick of jelly to spice it up. You know darn well where we get it. We can't live in fear. Eh, you want some ice. I know lots of fellas from the city. I'd mind my manners, seeing I'm an ex-cop! I know people who will sort you out. You see, there's a beauty in trying to solve crime. Every intimate moment of the action has to have a place on the crime board, like chess. The evidence has to make the right shape, you got to track down leads, make a picture, find all sorts of possibilities, the giveaway gaffs. It's got to have sex appeal to get city cops going. Lots of breasts and nice tight pussies will do it. But this one? Well, it's just a daggy kid with a drug problem. Too many of them. Not nasty enough, is it? No fuckin' suspense to get your dick stiffening. Needing a bit of lubrication. You got nothing here. Just some Mars Bar packets and a greasy sausage roll on your desk.'

Blackett gave an 'up yours' to Bam Bam and slammed the door. He felt a tremor in his heart; the stress was killing him. But he couldn't help himself. He opened the door to the interview room. 'Don't even think about leaving town!' he yelled, then looked at Johnson and said, 'Let them go.'

For once, Blackett wanted to be a hero, to solve the case himself, to have someone look up to him. He was seething with anger. At anyone.

CHAPTER 27

9th August

The day arrived for the Save the River community meeting, it was the International Day of the World's Indigenous People and the family were worried. The meeting might stir up trouble. The Wilga Community Hall was a brick and fibro structure 'up town', on the edge of town. No insulation and the heater on the wall was broken. Everyone wore hoodies or beanies and Ugg boots. The small houses nearby were neat and had little rose gardens and children's play equipment scattered among lounge chairs. The Uniting Church pastor lived nearby, and he kept an eye on everyone. Along the street were beautiful wattle trees that burst into yellow flowers in winter. Kids played and threw stones along the road.

June observed all the main players. Steve leant on the wall, breathing hard. He looked cold to the bone. She patted his arm. He was not naturally outspoken, but he wanted to offer his knowledge to the local Land Council. June thought he would make a future Black politician. For now, it was torture for him to speak in public. He wanted to not shake when he spoke up. Someone would call him out for not being a local, that it was not his Country. 'Don't speak for us, you're not from here! The water is our blood!'

Insects whirred through the windows; moths flocked to the lights. It was the end of what passed for winter. Steve sat next to June and smiled as he looked at Arana. She would give him strength.

Steve had told June: he saw himself as a ten-year-old lined up at the Queensland boarding school in a woollen uniform. English school clothes with silly little caps. The teachers recognised his potential and helped him win a scholarship to university.

Jack sat in the audience, and he was as good as gold, with not a breath of grog on him. The old people were occupying seats at the front; even Elders from out of town sat tall.

June moved to the stage with Merle, Steve, William and a few others who wanted to talk up, like Bundi. He reached over and squeezed her hand. He wore a blue cowboy shirt and an American Indigenous necklace.

June's heart beat fast and she held her fist to her mouth; she had a tendency to get nervy if she was at a meeting, but she needed to speak up. She felt guilty about not finding enough about Thommo's death.

On the wall was a nice gold-framed picture of Queen Elizabeth and the Australian flag. Jesus, she disliked that bloody flag with its red, white and blue. Slaves to the British Empire. June preferred to see the Aboriginal flag, currently draped on a lectern.

The hall was a shared space, even though the whitefellas didn't want the Murries to use it. They had paid for the night, so no one could stop them. Old coloured posters hung on a wall. Someone left them pinned from thirty years ago; they advertised the National Aborigines Day Talent Quest. There were a few white people in the crowd, like Doctor Ian and the medical pathologist's family. They waved at William and Merle, showing support.

Bundi gave a speech on the stage, and he paced back and forth and jumped down to stand before the audience.

June hid her face in her hands.

'In the last sixty years our river has started to die,' said Bundi, 'and now the unthinkable has happened and it doesn't flow. It's like a horror film, with baddies driving huge earth-moving machines. Water dribbles onto millions of cotton plants.'

'The bloody Murray–Darling Authority told us we had to buy water rights!' shouted William.

'We Aboriginal people need to stand and be counted, not quiver in fear of the white man. We don't need government to give handouts,' said Bundi. 'They divide our loyalties and pull us into their system, which oppresses us.' He raised his fist, and the place erupted to whistling and cheers. 'We are able to support ourselves if we have our own water and our own land rights, water cultural rights.' Bundi sat down.

Merle nudged June.

'Eh, who's here – it's Wayne,' whispered William. He had spotted the young constable standing awkwardly at the back with Sergeant Blackett.

Two elderly farmers from Menindee stood on the podium and held a blown-up picture of them holding enormous dead fish. 'Along forty kilometres of the Darling, algae bloom from not enough water has killed one million fish. Ancient hundred-year-old Murray Cod have perished, not because of drought but because of the actions of the federal and state governments, Big Cotton farmers and the useless Murray–Darling Basin Authority. It's not bloody drought, it's a bloody disgrace. We want to stand with Indigenous people and demand we bring these catastrophic events to the world!' The men were applauded, and they sat back in the audience.

June stood up as Merle took the microphone. 'We've for a long time listened to you men, now us women will speak,' Merle said. 'Here's our own private investigator.'

June unfolded her printed speech and stood up. 'We're here for one thing, and it must unite us. Black, white, Asian, gay, straight, trans, greenies, ferals, feminists, Labor Party, Shooters, Independents and even the National Party. Humans live in the age of the Anthropocene,' said June. 'A time of a great extinction of species. Humans make climate change. Not God, not farmers or irrigators or logging companies alone. Every day, around the world, they murder defenders of the environment to protect business interests. Like here in Wilga, a young man probably got killed for trying to protect our river, with corrupt police covering it up for whatever reasons.'

From up the back, Blackett yelled, 'That sounds like slander, lady!' He shook his fist, but the audience turned away from him and Johnson hid his face in his hands.

'The international panel,' June continued, 'what you call it, the Intergovernmental Panel on Climate Change, says that we have ten years to save this planet. How do we save ourselves? Our river? Who and what will be left after our actions?'

'Sounds like incitement to riot – that's an offence!' shouted Blackett.

'Let's not wake up and find we have let them down and there is nothing here on Earth but a desert full of landfill garbage and not a drop of clean water to drink.' The room was silent, then Arana stood and began to clap. The whole room stood.

Bob, a Murri man with a strong face and greying dreadlocks, walked to the stage. He took the microphone. The room hushed. He was from the city, but he could talk up – he'd been on the TV. A circle of white ochre ran around his eyes. 'I have lived all my life on this river and in my town it is dead. Yaamangunna Barka. We got sacred ochre pits for ceremony. We go out and work at mending our ancient fish traps. River is life. Every human needs rivers to sustain us. We call on all people to come to dance and sing for our land. Talk to the land; call on Biamee, the great spirit. Natji Waaway rainbow snake made this river. The Elders summon the Noongah, the Yorta Yorta, the Koories, Murries, all the people who care, to come and fight for our rivers. Droughts and man-made destruction make the river die. How do us lot bring back trees, birds, the fishes, the animals? We can't live without beloved Mother Earth; trees give us oxygen. We can't keep clearing land for cattle, we need trees. We're all suffering.' He grasped the microphone. The hall was alive with clapping. 'We relied on the river Dreamings and totems like the Murray Cod. Totems die, then we die. How do we live when the animals become extinct? They're tapping into that artesian basin; it got a lot of salt, lot of sodium. Not good for drinking. You ask family in Wilcannia, Dubbo, Bourke, Moree, all drinking bore water. Poison stuff. Now we see that migration to big towns. We have to try to fix it. First Nation people need to be part of those discussions and a big part of decision making. We need tanks; we need filters on the water. The political parties always sell out our river. seventy per cent is controlled by overseas companies. Cotton should not be grown around here. Rice, almonds, walnuts all take too much water. It's about our children's children. Rain-makers sing the river and land and we can bring it back but it's also about fighting for restorative justice. You are on a journey, whether you like it or not.' He sat down, and everyone cheered.

Old Mary stood up with great difficulty, and her walking stick shook. 'We must protect them places even more. The Nguma Malyan is the Valley of the Eagle near Bennetts Gorge. Special places for that ceremony and stone and ochre paintings, you know long time, olden time. Them places are Ngiyampaa and Barkindji country. My family got moved by trucks after that big Depression in 1933. We went to Carowra and then Menindee. We haven't even been able to protect our burial places, our grave sites.'

Some noise began at the back as a few white authorities arrived at the hall. Jack's knees bounced up and down and June put her finger up to her lips to quieten him. Her heart pumped. June saw Wayne Johnson push through the crowd of people.

At that moment, they heard a booming voice: 'Let me pass! I demand to speak.' Wilga Shire Council Chairman Mr Wright forced his way in, pushing people away as he moved towards the stage. He nodded politely to the old ladies.

'That you, Granny Mary? Feeling better after your operation?'

'Oh, look out,' said Merle.

June ducked her head, because the arrival of Warren Wright could make a stir. He was a well-known gin jockey, but also a man who supported all the cotton growers, even though he had Murri blood.

Sergeant Blackett cleared a path for the shire council chairman. He moved with power; his blue lapelled shirt stretched across his huge girth.

He faced the people on the stage, his back to the audience. 'I'm Chairman Wright, and I respectfully ask for a moment of your time. And please, young ones, don't film me, I haven't got me makeup on.'

June reckoned he needed drama lessons.

Constable Johnson blocked the entrance with his arms folded.

William stared at Wayne and shook his head.

Wright took a land rights poster from the table and held it up. 'You already have rights. We support you! I would like to acknowledge Elders past and present.'

William's fist clenched, and June put her hand over it as Wright turned and grabbed the microphone.

He poked his finger at the crowd. 'Look here, you are being brainwashed by out-of-towners. Fair suck of the sauce bottle. It's fake news, it's leftie indoctrination. It's not rational; you got to read the facts. You can all read, can't you? Plenty of water. It's just the drought! The council can't make it rain. That young man was trying to destroy the town's livelihood. Good thing he's no longer with us. Farmers suffer alongside you; we feel the pain. We can work together; you can give us a rain dance!'

'The river's gone!' shouted William.

'We are not running out of water. Reduced emissions mean what – that we won't be able to fart as much? Or eat beef? It's those climate change zealots who stir up trouble. This is a nice quiet town of fibro and tin people, and we don't appreciate Black activists. Or Black investigators. What do you reckon, Sergeant?' said Wright.

'The police stand for law and order,' said Blackett. 'We will protect property. And Aboriginal rights. It's the law.'

'You all cleaning your ears out for this?' said Wright.

'See the dead fish piling up?' shouted William. 'Can you smell it? It's not a river, it's a sewer.'

'Now, let's all be civilised, shall we? Keep calm!' said Wright.

'We'll dump a truckload of fish outside your fucking posh house, in your pool!' said Jack.

'You'd be charged,' sneered Wright. 'The government can't make it rain, you moron! Climate change is crap. We had the third-largest cotton crop on record. You should all be proud! The authority drained the Menindee Lakes to save evaporation. It's science!'

Wright turned to stare at June and Merle.

William itched to punch him bang on his beak.

The young people in the audience began to chant. 'Get off, get him off!'

Wright pointed at the university-educated ferals with long mullets lined up cross-legged on the floor. 'And you do-gooder hippy greenies from down south!'

'This is a private meeting!' said Bundi as he stood up with remarkable poise.

Wright nodded at him. 'Surprised to see you here, mate. You meeting us tomorrow?' he said.

Bundi took the microphone and began moving Wright from the stage.

Wright struggled. 'Just let me, for Christ's sake … Let me speak!'

'Get a straitjacket!' yelled a young man.

Wright's face was red as he pointed his fingers like a gun at the crowd. 'You're not getting any precious water rights unless you've got tons of dollars, and you're not having any more land rights!

We own the land, you just walked on it. Not getting cultural flow water now, not in ten years' time, not ever! Screw the lot of you, boongs!' His mouth frothed, and hell broke out around him.

'We'll see about that! We got water rights!' yelled old Mary, sitting bolt upright with great dignity.

Wright's fist clenched, and he slammed it against his other palm. 'They'll burn your houses down first.'

'They don't own any houses,' shouted Merle.

Wright pointed at Steve. 'You, we've got your name, fella. Allunga! We know all about you agitating with environmentalists and outsider city Blacks. Talking about sacred rivers! We'll fix you, all right. You'll wish you stayed in Sydney with your student radicals. I'm in the Shooting Rights Party, and we like to use Greenies' posters for target practice.'

'The fish are frigging all dead! You like eating carp, do ya?' shouted Jack.

'Sacred, my arse,' replied Wright. 'It's our water, our tourism money, my tour buses, the town's life blood. You're not going to destroy our businesses. We are building a monument to the explorers Burke and Wills and putting up a stockman's hall of fame.'

'I'll show you what I think!' snarled William.

Wright clenched his fists and raised them. William faced him with fists up, but Merle pulled him back to his seat.

'Stay there, old fella, and shut up!' Wright said.

'They're not telling us to shut up at our own meeting. You can't push us around!' yelled William.

Dribbles of perspiration ran down Wright's nose.

Old Mary moved out of her chair, and with great dignity walked from her seat. She moved to the back door and the police officer nodded at Wright and let her pass.

A few minutes later, Johnson stood in the passageway behind Arana. He observed her sitting with her flag rolled in her lap. She was magnificent. A shout rang out from the street, a commotion, and the lights went out. Someone called out to find the fuse box.

Johnson moved fast, leaning over an empty seat in the dark. He grabbed Arana's shoulder, pulled her towards him, and kissed her deeply on the mouth. She elbowed him in the chest. The lights came back with a bang. Wayne smiled at Arana.

'What did you do that for?' she asked.

'Huh? Wasn't me.' He crossed his arms, and with the stance of a policeman he moved towards the back door to stop the crowd rushing.

The noise subsided and Merle unfurled a pile of papers. She put on her best teacher voice, nice and clear and loud. She took the microphone from Wright. 'We are about to read the minutes, and William will give a treasurer report. Mr Wright, you need to take your place in the audience.' She unfolded the agenda. She could talk up when she needed to.

Wright walked back and forth in front of the stage. He leant towards William again and spit flew from his mouth. His huge body was full of beef, full of shit, full of high-status stink. 'I'll send a group of my blokes round to your Reserve to beat the living shit out of you. And if you go near the water operations like that fool Thommo they will lock you bloody up! Go on, Sergeant, do it!' Wright hissed but Blackett helped him move away from the stage.

Wright was having trouble breathing as he hung onto Sergeant Blackett, and they walked out of the hall.

Johnson slithered out behind them, looking miserable. He waved at Arana, but she turned her head and looked down. She didn't want to recognise him.

Merle called order and read the dull minutes.

Then William stood up. He sweated and looked suddenly shifty. 'The treasurer's report is short. Nothing much to say. We have spent the council grant on this hall and stuff. We're broke, but I have ideas about raising funds and helping to get a big group from the city to do a corroboree along the river with us to bring attention to the problems.' He sat down.

Someone shouted, 'Where's the money we raised?'

'Is that it?' said Merle.

'You can examine the accounts anytime you like. All above board, my maths is not that good, so I did my best,' said William.

'Who took the money?' yelled a member of the audience. 'Saw ya at the TAB, big-noting yourself.'

William stood up and faced the jeering. 'I told you, we had to hire stuff. Aunty Mary has more info.' He hunched back in his chair.

'We'll work it out, cuz,' said Bundi.

Arana stepped up to the podium and took the microphone. Her voice shook. 'We young ones are told to stay calm and study, when we want to shout like the young European activists: "The house is on fire … Do something!" The planet is burning up and politicians are worried about the downturn in income from fossil fuels like gas. We want to tell you: get out of our way! Recognise that all the species will be extinct, including humans! We young ones want you to panic!' She stood still, then the room erupted in applause.

They had a tea break and ate some scones and drank soft drinks; then it was time for Steve.

Steve was about to address the meeting; the Aboriginal people sat down with the tumult of children, grannies and youngsters. Young fellas on bicycles rode up and down outside, and some children kicked a football inside the door until someone told them to stop. By the back door ladies from the Regional Women's Association were selling cakes; they waved at June. She waved back. They held up a sign that said 'Save our river, save lives.' They also were selling booklets on Aboriginal massacre sites around Australia and had an Aboriginal design-printed cloth over their table.

June watched Arana handing her Aboriginal flag to young Steve. She grinned with pride. The family couldn't stop that kind of young love, as Arana was so wilful and would have her way no matter what William said.

Merle and June helped some Elders back into their seats, and shushed the kids. They had to learn respect when it came to a formal meeting, and everyone was proud to recognise this young fella who was about to speak. But first that fiery Bundi wanted another go at the microphone, and young Steve waved him up to the stage.

'You fellas have got to get up off your arses and fight for this river system,' said Bundi. 'We are not going to get ahead by complaining that the white man will never help us.'

'I hear you sold us out and joined the gubbahs!' yelled a fella in the back.

'I am only First Nation person associated with the Murray–Darling Basin Authority. I don't have the mandate to represent thirty First Nations.'

'We got algae bloom killing all the fish!'

There was a hush. Bam Bam walked into the hall, carrying his bike helmet. He sat near the door and winked at Duckie, who was slouched against the wall. She froze and hid her face.

Bam Bam slowly stood up and held up his hand. 'I've got five thousand dollars to donate to the Save the River fund.' He waved a pile of bank notes. 'I would like to hand it over to Aunty June, as long as she makes a public apology in the *Wilga Gazette*. My reputation has been besmirched around town; I might be part of a motorcycle club, but that doesn't make me a criminal!'

'Good on you, fella!' shouted a Murri youth.

'Okay, now, let the young ones speak up!' said William.

June ushered Steve towards the stage and Arana stood at the side, glowing with pride.

'Thanks for coming, everyone.' Steve stood tall. 'I'm here from the New South Wales Environment Defence Group. It's great to see so much community interest in the cultural flow of the Darling River. No water is flowing over the weir. We need to end water being channelled to the big irrigators. In the past, Murri people have agreed to work with the Murray–Darling Water Authority or New South Wales Water. But you've been conned and sold out. Some of your own have sold out! Lots of meetings with them giving acknowledgement of Country and

Elders and all that, but it's just nice fashionable words! No real action. They said you could buy back water for cultural flows. It will be millions.'

'The bloody shitheads!' shouted Jack.

'The river is sold to international consortiums. It dries the Dreaming tracks out, genocide for our spirit.'

Bundi stood up on his chair with a sign: 'We demand water rights for culture'.

Steve continued. 'The Gwydir district farmers have just bought millions of gigalitres that was meant to go to the river.'

The room was deathly quiet.

'They have robbed all people – Black and white. And the big landowners just get richer and richer, and activists just disappear. We invoke the Rule of International Law that we have never surrendered nor acquiesced in our claim to these territories of land and water. We'll go to the United Nations for our rights!' Steve stood next to Bundi, and they linked arms.

It shocked people when Steve raised his fist in a Black Power salute. William and Jack nearly fell off their chairs and put their hands over their faces. June stuck her arm in the air.

'Well, that's about it,' said Steve. 'We will be working with a film crew to record the river theft up near Brewarrina. Watch out for the exposé on the TV. Now, some Elders want to have a word about an old cemetery. Anyone wants to talk to me, I'll be out back.' He climbed down into the audience and sat next to Arana.

After the meeting, June watched William talking behind the hall to Blackett. They were laughing at something, then Blackett hunched over and William handed him an envelope, which was slipped into his lapel pocket and patted. June screwed up her face, and confusion caused her to slip on the pebbles.

'Careful, Aunty, hold on to me,' said Arana.

Back at the house, June confronted her brother. 'What's going on with you and Blackett?'

William looked at the floor and mumbled.

'Tell me.'

'Nothin'.'

'Tell me, darn you!'

'He wanted to read some of my poems, and I gave him a copy.'

'Shit, I wish I believed that.'

***

Next day June kept searching for Thommo's killer. She navigated to the National Socialist website and registered as a potential member. Anything to get information. However, she was soon kicked off the site when her answers to some questions were incorrect.

She asked every person she knew if they had seen any suspicious men in town around the day of the murder. Only one person said he thought that a stranger was following Thommo, speeding through town in a white ute and tailgating him. No name to be had, and no evidence about anything to do with Thommo's end. June thought she might let the police know, anyway. Someone might have reported it.

CHAPTER 28

11th August

The next day, Steve was on the track leading to the river. It was a warm morning, the end of winter, with dust flying through the air. His white t-shirt had environmental logos. He tossed his head, and the sun caught the curls and lit up his head like a Renaissance painting. The tall pinkish coolabah trees shone in the early light, and the river trickled as a brown rivulet. He stood next to a television producer, who wore her hair in a pink and blonde flip. Janny was an attractive Sydney-based journalist, and she waited while the sound operator adjusted the microphone for Steve. Her false eyelashes looked like little animals. He was nervous, but realised he had to speak up – after all, he'd brought the TV crew to town. They held the microphone over his head and Janny moved close to him and touched his bare arm with long silver fingernails.

'Here we are in the outback,' she said, 'on the Darling River, with Steve Allunga, who will tell us about the loss of this water and the mysterious disappearance of a local activist.'

'There are certain irrigators disregarding recent government reports and illegally taking huge amounts of water for private use,' said Steve. 'Some cotton farms are like great inland seas. This water belongs in the river system, to Australian people and to us Indigenous peoples.'

Sounds in the distance echoed around them: a bulldozer was tearing into bushland, and roaring trees came crashing down.

'Go on,' she said.

'Most irrigators are along the Barwon, flowing into the Darling. We allege certain New South Wales government bureaucrats turn a blind eye to the overuse and illegal practice,' said Steve.

The camera whirred.

'Can you name them?'

Steve smiled and shook his head. 'Not at this time, due to legal reasons, but they will get charged for theft. It's the same as stealing money in a bank hold-up.'

Janny waited for the camera and a close-up while a flock of loud cockatoos flew over their heads. They waited for the cacophony to end. She liked this handsome young man, and smiled with an inviting head tilt. But he did not respond to the flirting; he was a man on a mission.

'There is no question the majority of Australians support the expenditure of 13 billion dollars of taxpayers' money to fix this river,' said Steve.

Suddenly, a huge white truck burst out of a small bush track. A thin, rugged man in a pink high-vis vest pulled the truck over next to the TV truck. Another man climbed out – it was a guard, wearing a red baseball cap tipped backwards. Sweat darkened his t-shirt with 'Gobblers' written on it.

He rushed towards them with his fist up. 'What are you doing on this land?! It's private!' he shouted.

'We're on the sealed road, a public right of way,' said Janny.

'This is a consortium property, and you'll buzz off or we'll get the police,' replied the security man.

'We'll leave when we've finished. We have rights too, sir.'

'I'll see about that,' he said as he dialled his satellite phone and climbed back into his truck.

'It's legal,' said Janny.

'You hippy types. Just follow the stink of dope and patchouli oil. You're a lot of do-gooders from down south. Try waiting for the drought to break.'

The air was thick with suspicion, and Steve took a photo of the truck with his phone. The man in the driver's seat ducked down and stuck a finger up to show contempt.

Janny turned her back on the man and gestured to the cameraman to keep filming. She smiled and Steve faced the camera.

'A lot of irrigators got a huge windfall out of this,' she said. 'Do you want them to put water back and make it flow?'

'It seems to be a knee-jerk reaction with a bailout,' said Steve. 'The Murray–Darling Basin's survival hangs on a desperate 12.9 billion dollar injection. It's a seventh of the continent. Our Indigenous Elders demand water back to allow native fish to breed and to protect cultural sites and ancient river red gums. It's a water source for three million people, and the life for the internationally recognised wetlands Lake Menindee.'

The man from the truck came lumbering towards them, shaking his mobile phone. 'I talked to my supervisor. Reckons you're trespassing, so he's coming out with coppers to arrest you. We have rights to protect this legal business.' He shoved the camera operator.

'Hey, keep your hands off,' said Steve.

'Our private security are all Samoan karate experts. They're on the way.'

'Butterballs, eh?'

'We are in our rights—' said Janny before the man snatched at her script.

Steve stepped forward, fists raised, but Janny pushed them down.

'Have a go, you rock ape!' The truck man stuck out his chin.

Janny laughed.

'You'll be laughing on the other side of your pretty face in the lock-up!' said the man. 'We've got you on film, mate.'

'We'll be moving along in a few minutes. Just please let us film,' said Janny.

The truck man hesitated and stepped back, leaning against a tree. Janny motioned to keep recording.

A motorbike roared into sight, and a large, plump, black-shirted security guard in riding boots hopped off his bike and strode towards them.

'Not another idiot,' said Steve.

'Don't approach us, sir,' said Janny. 'We are finishing a national broadcast on TV, and you'll be in shot.'

The young Samoan stopped and regarded the scene. He had a knuckle-duster clenched in his fist, but he stopped and gazed at Steve. 'Really? Can I say a cheerio to my mum in Bundaberg?' He waved at the camera.

'Confiscate their shitty gear! It's trespassing. Do your job!' said the truck man, and he tried to take hold of the camera.

The operator swung a punch as Janny jumped in to separate them.

The Samoan pushed them apart, then recognised his mate, Steve. 'Hey, I've met you. Steve, eh? I know your mob up home. How good is Queensland? Good to see you, bro,' said the Samoan.

'Yeah, let us do our job, eh? Saving the planet,' said Steve.

'No worries, bro,' the security guard whispered. 'Hey, you know Cowboy? He's my cousin. He was bragging about stuff, a hit on some greenie.' He laughed and rode off.

The truck man was speechless. He climbed into his truck to wait for back-up. Steve then recognised another companion in the truck. It was Bundi, his hat pulled low and head turned away. The camera whirred.

Janny continued. 'We understand the government is buying back water from farmers to give the environment a drink, that sounds uncontroversial. What do you say?' She looked earnestly into the camera before flinching at loud cockatoos.

'Yeah,' said Steve as he wiped his face of sweat. 'Many farmers reckon the government is wasting big money buying what could be very productive land and turning it into bush again.'

'Does it cause big arguments out here?' asked Janny.

'Global warming is destroying much of the land, and creating wild bushfires, and turning it to dust. We owe it to future generations to fight for Country, otherwise it will be dead earth. Like Mars.' Steve surveyed the devastated riverbank.

Janny told the cameraman to shoot the piles of collapsed coolabah trees that were clogging the scrap of brown river.

The journalist walked back to Steve as he drank from a water bottle. She told him that she had explosive footage showing direct water theft with broken measuring gauges along the river, pumps going night and day. They had received immediate threats from Big Cotton farm workers who were recorded. The TV episode was going to be dynamite.

Steve spoke directly into the camera. 'The government's water buyback is causing a water war. There are angry divisions between north and south. The Coorong River in South Australia is dying because of greed. Rich cotton kings protected by National Party ministers. It's a crime! Northern New South Wales landholders point north towards bigger water users upriver. "Send water south" is the demand. Water stolen in broad daylight and irrigators paid millions in government buybacks. If this was white-collar crime in the city they'd lock them up.'

'What other issues impact on local Aboriginal people?' asked Janny.

At that moment Bundi walked into the shot and moved Steve out of the way so he could speak to camera. 'We need to speak out about Indigenous incarceration rates,' said Bundi.

'How about your mates,' Steve said, 'the cotton farmers stealing water, and nothing happens? What are *you* doing, Bundi? You're rubber-stamping government policy.' He pushed his chest forward towards Bundi as the camera kept recording.

Janny grinned. This was a drama.

Bundi fronted right back. 'You want a go? I spent years learning to fight little shits like you in prison.' Bundi's face pressed up against Steve's. The tension bristled between the men.

Bundi continued. 'And William is in debt up to his ears for gambling, that's why he took the group's funds. Sort that kind of corruption out! He even owes Wright and Blackett money.'

'That'll do it,' Janny said and pushed them apart. 'This is not WWE wrestling.'

Bundi winked at her. 'If you give it a go, you'll get a go, eh?'

'You can leave, Mr Bundi,' said Janny.

'You want to meet me for a drink later, sweetheart?' he whispered, then laughed and moved towards the truck.

Janny watched him go. She called out: 'You might be an Elder, but you are a sexist arsehole!'

He turned back. 'I'd like a go at your sweet hole. After a pinot noir in your Potts Point apartment!' Bundi roared with laughter, gave a little skip and wiggled his shoulders in corroboree movement as he climbed into the truck cabin.

The journalist shook Steve's hand, sprayed her face with Evian water and shaded herself under a white umbrella. She reapplied makeup near a tree and swatted flies. She put on a bush hat with green fly gauze. The film crew sat companionably in the shade and ate sandwiches. They ignored the truck with Bundi and the high-vis man. The crew didn't offer them a smoked salmon sandwich.

Steve noticed the sound of a far-off police siren approaching, so he hurried the crew into their vehicle and drove off in a blast of dust, giving a two-finger salute to the men in the truck.

So now he knew Bundi was a traitor to the river.

* * *

Back at the house in Wilga, June and William sat outside at twilight when Bundi drove up. He climbed out of his vehicle and sat on a milk crate out the back next to William. June moved and closed the back door with a slam.

Bundi knocked gently. 'Come on out, I won't bite. Peace!' he said.

June shrugged and walked back into the garden.

'Haven't had a call lately, eh?' said June.

'Come on, cuz,' Bundi said. 'I been busy making Aboriginal people self-sufficient with business enterprises. I want to convince our people about the economic positives in working for Big Cotton. Can't beat 'em, join 'em. They've got a position for you too, William,' said Bundi.

'Listen to yourself,' said June.

'He might be right. We got to get independent of welfare,' said William.

'If the river is gone?'

'You can get a real good water purifier. I got one,' said Bundi. 'Uses lasers to clean the water. In a few years the government will build a saltwater desalination plant up near Brisbane. Pump us all the water we want right out of our tap.'

'How many years will that take?' said William.

'Only ten. It's got to get federal funding. June, you could move to Queensland. Forget this river country, it's finished. Lots of opportunities for your family. You could put Jack in a facility, you know, top-class. Visit him often. I can get you a job. But maybe working wouldn't suit you.'

'What's he saying about me?' yelled Jack from the back door.

'Maybe Bundi is right,' William said. 'It's all over. No river to fight for. It's a joke to go on with the struggle. Capitalism and global corporations have won. It's invasion and conquering all over again. We're failures.'

'Even the owls have packed up and left,' June said. 'I haven't heard one for weeks – nothing to eat. Marsupial mice almost extinct.' Tears stung her eyes. 'Might as well kill myself.'

'Don't be silly. Who's going to cook dinner? We got kangaroo sausages,' said William. 'Who's going to teach the young ones about our culture? The Dreaming stories. Keep hope alive. The young men are doing ceremony again. Corroboree on the oval yesterday, I seen it. Couldn't believe it at first. But I trust the young ones to bring us through this crisis. They've got strength, same as we did in the old days.'

'Getting late, eh?' June inclined her head to the street, and Bundi nodded.

'Off to bed eh? Sorry to upset you, June. Something to think about,' said Bundi.

'I'll never stop fighting for the river.'

'You got a bit of milk froth on your lips, just there.' He touched her bottom lip, but she pushed him away.

'It's not a rubber stamp. We need a First Nations representative on the Murray–Darling Basin Authority board. I'm it,' said Bundi, and he grinned at June. 'Look, I know I got no mandate to represent the thirty First Nations along the river system. Cultural environmental vandalism and water dispersal: we can fix it, but it will be twenty years.'

'Yep, it's late,' said William as he tipped over his milk crate.

'I got some chocolates in my car for you, June,' said Bundi.

'Better give them to your other women,' she said, and stood up.

He watched her nice backside as she walked to the door.

CHAPTER 29

20th August

The fans whirred in June's house. Winter was over. Some days she just wanted to forget about the town being a Peyton Place; forget the politics of environment and forget race and forget dead cousins. She sewed extra pieces onto her ball dress, ready for the gala night of the Debutante Ball. She tried on some black panties and a bra, just in case she got lucky. She pushed her bosom up in the mirror and winked at herself. Despite her efforts, she couldn't get Bundi out of her fantasies. She imagined him without his shirt, his rippling muscles and dark skin taut and luscious. June climbed into her bed and pressed her thighs together.

Early the next day, Merle cleaned shoes out in the shed and William mowed the lawn even though it was turning brown. June rubbed her eyes at the breakfast table, and Merle patted her cheek and poured tea.

'Here's your toast with Vegemite.' Merle smiled.

'Hey, June, cold last night but today it'll be the warmest day for end of winter ever recorded, eh?' William yawned.

'Yeah, hit thirty degrees here yesterday,' she said. 'Nearly died of exhaustion putting up the ball decorations. But we made the hall lovely. I love the footy and the deb ball.'

The Knockout results pleased the family: Walgett had won again. The Wilga mob had pride in their cousins, and now were ready to have a good time. Team supporters drove around town yelling, with flags waving, celebrating. The football made everyone happy. Now it was time for the ball, with dancing, flirting, eating and finding true love. A ball attended by Black and white people, bringing everyone together.

By six o'clock the fellas were dressed up in suits from St Vincent de Paul, and they sat in the kitchen drinking coffee; they ate Jatz biscuits and June could hear them laughing and telling jokes as usual. They pulled at their ties as if choking.

Merle and June dressed for the ball in the bedroom. They laid out bright colourful dresses with low necklines. They had bought them second-hand, but they looked very nice.

'Show off those nyamus? You got a spare push-up bra?' said June while she adjusted her bosom.

'Shame job, I'm not stuffing tissues down there.'

'Now, you want to keep an eye out for that Betty! She's got a hot moot. She's after William,' said June.

Merle combed her hair into a bob and sprayed it with hairspray. 'I trust him,' she said.

June applied mascara to her lustrous eyelashes. 'You can never be too sure. BB: all box and boobs. I reckon she's got munyi in her hair; always scratching.'

'She comes over, she's gone,' said Merle as she sprayed perfume over her wrists.

June smoothed her red ball dress. She imagined trying to wriggle out of it later if she found a nice fella. 'I'm getting desperate. I might doori that old Blackett. Give him a knee-trembler up against the police station desk. He likes Black ones,' said June.

'You're terrible. Better breathe in for this. You're filling out, June. Taffeta doesn't stretch.' She zipped June up.

'The men want a bit to hold on to.'

There was a tapping on the front door, and Jack opened it to find a dishevelled Duckie with a bowed head.

'I want to talk to June,' she said, and pushed past Jack into the bedroom. 'Hey, Aunty, please tell me why nobody seems to care about Thommo, but! Not even his mob?'

June sighed and sat Duckie down on the bed and held her hand. 'Maybe some blamed him for stuff. Hold a grudge, about the events in the past.'

'What stuff?'

'He didn't show enough respect for some people.' June handed her a tissue and patted her back.

'Can you find out who did it? I need to know. My boy will want me to explain, you know, why he has no dad.'

'Come back tomorrow, we got to go out,' Merle said, shaking her head.

'There might be information you don't want to hear,' June said.

'I don't care what he done, but,' Duckie said, 'I just need to understand so I can accept it, but. I'm begging you to find out who murdered him!'

Jack stood by the door and nodded his head. 'I bet I know who killed him,' he said. 'It was the old sergeant, 'cause he wanted more of the profits from dealing yarndi.'

'Don't talk about things you know nothing about,' June said, and shook her finger at him.

'You tell me, Uncle,' said Duckie.

'I saw it in a dream,' whispered Jack.

June asked Duckie if she wanted to come to the ball. There was a nod, and Merle looked in her wardrobe for a dress. They decked her out in a pink retro eighties dress with puff sleeves, and found a pair of Arana's shoes to fit.

June combed Duckie's hair up into a bun and pinned flowers in it. 'You look real nice.'

Duckie's tattoos stood out against the satin. She pointed her toe and laughed, then paused. 'That's the first time I've laughed since it all happened. I look like shit, but thanks anyway. I'm only going so I can show my face, but. And I want to listen to any gossip about my man.'

June stood by the door and looked at the boys sitting in the kitchen looking smart and not drinking. 'Them fellas are doo-dahed up in their suits. Even Jack looks good.'

'Let me get a go at that mirror. I'm not used to this makeup – shame, eh?' said Merle.

'Pass me that scent,' June said. 'I'm going to wiggle my moot tonight, all right. There'll be plenty of spare fellas down for Knockout.'

'Eh, you got any better shoes than these? I can't wear high heels, they're killing me,' said Merle.

Arana walked in from another bedroom. White satin and lace. Her radiant face shone, and her hair was piled on her head with tiny flowers, and her long white gown hung from her shoulders, revealing dark smooth skin. Her diamante earrings shone like stars.

'Did the dress turn out okay?' she asked.

'Oh, dort, you're absolutely beautiful!' said Merle as her breath caught.

'You'll get mobbed,' June said.

'I only want Steve.'

'Always good to have a few spares.' June laughed.

Merle sighed, and they both stood next to Arana in the mirror to admire her. Only seventeen, but so lovely and so intelligent. The women were afraid for her. It might be better if she was less attractive. It caused too much trouble. How to protect her from men drooling after her like randy dogs.

They sprayed their hair one more time, and Arana stopped and whispered, 'I talked to that ice dealer about Thommo, but he said I should mind my own business.' She swallowed and tears welled in her eyes. 'He wanted to know where you lived, Aunt. But I said nothing, and he sneered, reckoned he knew everything about your snooping detective stuff, and that you should watch out when getting in your car. That accidents happen. His eyes were cold grey dots.'

* * *

The Wilga Aboriginal Black and White Debutante Ball was the Murri event of the year, with everyone in new clothes and bringing sparkling wine and beer and having fun. They decorated the hall with black, red and yellow balloons, and the tables were set with white cloths and wildflower decorations. Plastic glasses next to plates of supper. Sausage rolls and tomato sauce, sandwiches, cheese cubes with gherkins and hot vol-au-vents. Young couples stood nervously at the side of the ballroom in black suits and gorgeous white formal deb dresses.

Jack, Merle and William danced while June surveyed the dancing couples and checked that no one was drinking too much. The local Murri band played country rock, and the compere was dressed in a black satin country-and-western suit. He mimed to an Elvis song, and Jack mimicked him.

'Go Jack, you deadly fella!'

The compere took the crackly microphone and spoke. 'Ladies and gentlemen, let's give a round of applause to the ladies committee for the great supper and their work on the decorations. All proceeds tonight go to the Save the River movement. We got to fight Big Cotton! Not fight each other, with whites against Murries for the little water we have left.' Everyone clapped. 'Now, let's hear it for the winning footy team. What a great game, eh? What do you reckon?' More applause. 'You need to be upstanding to toast the young couples. This is their big night. Lot of these young ladies have sewed their own fabulous gowns, and aren't they gorgeous? Ladies and gentlemen … to the debs and their partners!'

Everyone toasted the debutantes.'To the debs!'

Jack lurched towards women sitting at a table near June. 'How you going, girls? Don't mean to intrude, but am I your partner, or what?'

'Jack, get over here!' said June.

'Nah, there's two nice dubais over here. And I'm as hot as a hot dog. I don't muck around. You go for Walgett, don't ya? Well, ya ain't gunna win again next year. I'll come back, or what?'

'Jack!' June yelled.

He smiled at a pretty one. 'Later, out back, eh? You and me, we might go. She likes me, all right.' Then he winked at the white women at a nearby table and all the Murri women rolled their eyes. He pranced in long swooping strides around the hall. He looked crazy, but had beauty in his freedom, gliding like Fred Astaire.

The compere grabbed the microphone and gyrated his hips. 'The ladies sure are looking attractive tonight. Any of you single ladies want to come up and dance with me? Everyone, get out on the floor – they've got a bottle of champagne going to the best dancing couple here tonight.'

'Eh, Merle – he's a real black Elvis,' June said.

'Some young ladies can't find their partners. Come on, gentlemen, do the right thing! You making me shamed,' said the compere.

'Eh, all you Walgett mob,' shouted Jack. 'You only won because of the winger, dancing up the sideline like a bush turkey. You won by one little point. I would have stopped him if they'd let me have a go.'

'Only Knockout you were in was with the barman at the Exchange Hotel,' called William.

'You can't hold your drink, Jack,' June shouted over the rock music while dragging Jack into his seat. 'Gee, that footy player … What a neat little bunch he's got,' said June.

'Whenever he put the ball down to kick he'd bend over in front of us, and all the Wilga women went arrh!' said Merle.

Jack cavorted wildly, and Arana walked up to her mother's table. 'Mum, what're we gunna do? We shouldn't have brought Uncle. He's shaming us,' said Arana.

Merle shook her head. Everyone had a drunk uncle, it was normal.

Jack danced wildly.

'There he goes, swinging those hips. Watch out!' June said.

At that moment, a hush as the didgeridoo yidaki echoed through the hall. An Elder walked ahead of five boys dressed in ceremonial ochre paint, red lap-laps and long feathered head-dresses of emu. They danced a goanna sequence, and two of the youngest deb partners, in black suits and bow ties, joined them on the floor. Their precise jerking manifested the hunters and prey of a hunting ceremony. The crowd cheered and stamped. Cameras on phones were lifted to record the Aboriginal moment of pride and identity.

Duckie drank constantly from a cask hidden in her lap, and eventually she lurched from the table as June tried to stop her. She strode up to Constable Johnson, who was doing a lame barn dance with one of the aunts. Duckie's tattooed arm reached up to his shoulder and she shoved him. 'Hey, you, copper, why haven't you found Thommo's killer?'

Everyone stopped and stared.

William took the microphone. 'Time for a bit of old rock and roll. Take your partners and we'll be giving a prize to the best couple.'

Johnson pulled out of Duckie's grasp. 'Sorry, miss, you've drunk too much. Might be better if you went home.'

'Home? Is that what you call it, but? A shitty shed out of town where I shiver in the cold with no husband, but. He was murdered! You've done nothing! I might be next. It's corruption, but. You taking a kickback on drug deals or something?' Then Duckie swung a punch at Johnson.

He called for help from June while holding Duckie at arm's length. Dancers swirled around them.

'You've done nothing, but,' Duckie cried. 'No charge, not one damn arrest, but. You're weak, but! Little copper with no guts. Lazy fat pig of a sergeant just sits on his arse or chases gins.' Duckie lifted her skirt and flapped it at Johnson.

'Language, miss!'

'I know you all think I'm a slag. Well, I'm not. I loved Thommo, but; he was the only one. Just 'cause they think they can gang-bang me anytime they got a gut full of booze.'

Johnson looked horrified; it was supposed to be his day off – at least until his evening shift. 'You need to help us lay charges,' he said. 'You ran off without finishing your statement; we were left with nothing.'

'I'm not that person who gets raped, but – never was, never will be. I'm not blabbing to cops. All you ever do is humiliate women! I don't fit into white society or among the Murries. Only one who ever help, but, was the investigator June Thomas and the women's refuge.'

June held out her hand, and Duckie took it and they exited the hall among sighs of relief.

Poor Constable Johnson collapsed back on his plastic seat with his head in his hands.

A few minutes later, June came back and shrugged at Merle. 'I got an aunt to drive her home.'

A minute later, Steve arrived in the hall in a dinner suit with a silk bow tie. He looked city-smart. Arana's eyes took him in, and the women sighed.

'Eh, Steve!' shouted Jack above the music. 'Come over here. You missed that kick, eh, brud, big shame job, eh? Bet the team jammed it up you.' He shook Steve's hand.

'Don't mind me, Uncle Jack, I'm trying to forget it.'

'Eh, brud, don't dance with those dubais over there.' He pointed to the women as they bent under the table, stuffing mouths with handkerchiefs. 'They're too flash, those uptown gins. I reckon they got Big Cotton boyfriends.'

Steve approached Arana's table and he smiled with warm love. 'Thanks for the warning, Jack. Think I'll ask these women over here for a dance. They look all right, what you reckon, Jack?'

'Lovely dubais.' Jack grinned at Merle and Arana.

'You're lovely in that suit, Steve,' Merle said.

'Reckon I look like a dickhead. I'll dance with the most attractive dubai in the place. Arana?'

'Don't take any mind of Uncle Jack,' Arana said.

Jack wiggled his Elvis imitation hips. 'Never mind me, I'm just a little Black boy,' said Jack.

'Aye, Jack, I'll buy you a Coke later. Come on, Arana,' said Steve as he swept her onto the dance floor.

'You should stick to shandies, Jack,' June said, patting him, and he danced off.

'Aye, don't you worry,' he said. 'I'm not the kind of blackfella that'll drink today, drink tomorrow. I want to get to the end of the line, you know.'

June danced with a fella in the jazz waltz. She leant against his suit coat. He was a good mover but she sat down after one dance. She wanted to keep alert for the young ones.

Bam Bam arrived at the dance, and he was looking over at her. He seemed unable to have worked out the dress code; he wore his bikie colours with a bow tie and a rose. A sense of trouble brewing would not go away.

He nodded at Bundi and asked June to dance: 'Like a turn around the floor, Aunt?'

'No thanks.' June turned her back, then with a change of impulse she stood up close to Bam Bam's beard. 'What are you doing here?'

'Dancing. Why don't you dance with your Black stud boyfriend over there?'

Steve and Arana stopped dancing and looked over the dance floor at June and Bam Bam.

'What's he saying?' said Arana.

'Want me to find out?'

'No, Aunty can look after herself, she's not scared of him.'

Still, Steve walked up behind Bam Bam and waved at June.

'Bugger off, pretty boy,' said Bam Bam. 'You know what happens to fellas that stick their nose in other people's business?'

'Get their heads bashed in?' said Steve.

'Watch it, smartarse.'

Steve moved back over to Arana. He held her close. They danced and leant against each other.

Across the hall swept Arthur, and he had his eye on June.

'Don't you dance with whitefellas?' said Arthur, and he sighed.

'All right, Arty, but no creepy stuff,' said June and she stood up.

'Can't guarantee your bosoms won't be on my mind.'

'Just keep them on your mind and not in your mouth,' said June. And she danced with him. He looked tall, fit and handsome, and his bald head shone. June inclined towards his neck; his aftershave made her swoon.

'I've been wanting to dance with you for years,' said Arthur.

'I'm not sleeping with you,' said June.

'You're a beautiful woman,' said Arthur.

'Your hand is on my backside, move it higher,' said June.

'That's what they all say.'

A big shape appeared over Arthur's shoulder. Bam Bam smiled.

'Mind if I cut in?' Bam Bam breathed in Arthur's face and took hold of June's hand.

'I do mind, I don't dance with potential rapists, thanks,' said June as Arthur swept her away.

'You'll keep!' shouted Bam Bam.

'I heard that William's got sticky fingers and the river fund ended up at the TAB. And he owes someone,' Arthur said.

'Right, thanks for the dance. I'm tired,' said June. She shoved Arthur away. Her head held up, she sat back at her table.

Arthur followed and whined, 'What did I say? Come on, sweetheart, I never meant to offend you. I take it back, June, please! I shouldn't have said that. Argh, June, give me a chance!'

It was no use, and he hung his head and trudged back to the bar and drank a whole can of VB in one gulp.

June whispered to Merle, as they had noticed Bundi waltzing in the middle of the hall with his occasional girlfriend, Betty. He was dressed in a snazzy city suit and his teeth looked newly capped, like an advertisement for whitening toothpaste. June hid her face while Merle raised her eyebrows. He exuded confidence; his eyes flickered towards June.

He glided his partner over in a jazz waltz. 'Why aren't you dubais dancing? Waiting for me?' he said.

'Nope,' hissed June.

'How's Granny Mary? Need any washing-up?' said Bundi with a wink. 'Catch you later?'

'In your dreams, fella,' said Merle.

Betty thumped him and he laughed, then took hold of her wrist, twisting it back, all the while smiling at June. Betty squealed, but shook her long hair as they swung around the dance floor. June could see a diamond ring on Betty's finger.

A few minutes later Bundi was back. He meandered over and sat beside June. 'There's consequences for our actions, eh Junie? I'll be back at Mary's tomorrow while she's in hospital. What you up to? I'm on fire.' He pushed back the chair and danced away.

June strode after him as he fled among the dancers. She tugged his coat.

He turned with a grin and looked at her standing with hands on hips. 'You got a problem, darling?' he said.

'I'm telling you straight: have a heart, forget about it. Try to be a decent fella!'

'Respect.' He sighed, and ran his finger down her neck.

'Don't,' she muttered. 'Just leave it; never happened.'

Some women nudged each other and sniggered.

Bundi leant towards June. He could see her blazing dark eyes, and maybe tears. 'Not a word,' he said as he zipped his lips and walked away.

June sat back with Merle, and her plastic chair tottered, so she sat up straight. 'Have I got a sign on my back tonight that says "Free doori"?'

'Could be worse?' said Merle.

'No, it couldn't,' sighed June.

'How about I write a Post-it note for you that says: "Moot closed"?'

Across the dance floor drifted the environmentalist, Effie. She was dressed in a dark men's dinner suit with tight pants and her hair was dyed purple. She was sexy and gorgeous. She ran her hand across June's bare arm. She smelt like wildflowers. 'Fancy finding you here without a partner. Would you like to dance with me?'

June smiled and took her hand, and they danced the cha-cha while Merle took photos on her phone.

'I don't give a damn what people think,' said June.

***

Up the back of the hall a group of young white men slouched on plastic chairs. They were clean-shaven, wearing riding boots, jeans and neatly ironed checked cotton shirts. It was a uniform. With tattoos and sunburnt arms. One had a Murri debutante partner, and he stood respectfully at her side. Other young men leant forward on the trestle table, drinking beer from plastic. They gazed longingly at the dancers. They suffered outback shyness. Two of them wore black t-shirts with suspect white insignia and had arms covered in satanic tattoos, and among them, with his head buried in his beer, was sad Bam Bam.

***

A lonely sight was the man with a white shirt and awkward clip-on bow tie. Wayne Johnson, incognito. In an agony of longing, he moved towards the chair where Arana sipped a ginger beer. He asked her to dance, but she refused him, and it hurt. She touched his arm while shaking her head. Wayne strolled away, all cool, and whistled, but slipped on the beer-covered floor. He pulled himself up and tried to regain his dignity in the face of people smirking at him. He stared at Arana as she swept by in Steve's arms. Wayne examined her every dance partner. He gritted his teeth. He would like to thump the young Murri who gripped her backside in the Pride of Erin. Could she see him up the back, peeking? His eyes never left her as he hid under his Akubra hat with a beating heart and sweaty hands.

They gulped the beers, and eventually the single men asked young women to dance. After all, the night had a theme: the Black and White Ball. Cow cockies' sons and town fellas all mixed up in the progressive barn dance. Even Bam Bam got up to dance; he had a gentle style with swinging hips. But rocked it out when the music changed.

Wayne drained his orange juice to the last drop. He had to be sober to start his shift. He was so totally lonely he might die. He thought he might ask William's sister, June, to dance. Or Merle, she seemed nice. He took a tentative step, then retreated, feeling ridiculous. People might stare, watch him making a fool of himself. Blokes on the street calling out: 'Black-lover.' Well, he was, sort of. He observed Arana with her hair rippling over her back, her head tilted in a laugh. Her long swanlike throat and her white dress cut low revealing a shapely back and shoulders. Oh God, he could be dying. She was so utterly lovely. He must get a grip on himself. Be cool.

Wayne had to be on duty at eleven. He had to go back to his pathetic room and make a Nescafé, really strong, then join the sergeant at the station. He had seen enough of other people having fun. He made a mental note to take dancing lessons one day. He had two left feet.

In spite of everything that divided him from her, he had a grievous longing that was reminiscent of a crush he'd had on a girl in third grade. She was dark too: Chinese and gorgeous.

Just as Wayne was leaving, Arana whispered to her partner Steve and walked past Wayne on her way to the ladies toilet. She nodded at Wayne and smiled, and he saw that she had a beauty spot on her cheek. Darker than the surrounding gold skin. It made him feel intense regret that he had not stayed on at school and gone to university.

The young cowboys at his table smirked and nudged each other as Wayne eyed the dancers.

'Ask her to dance again,' said Cowboy.

'Cluck, cluck, chicken. She's just a black slut,' hissed another.

Johnson stood and smoothed his hair. He spotted Arana on her way back to her table. He walked across the hall quickly to face Arana.

'Now, please, do you want to dance?' Johnson mumbled, and she nodded. He couldn't believe it.

'Just one,' laughed Arana.

Wayne took her hand and pressed her against his chest. She smelt fantastic. He swept her into a clumsy waltz and twirled her until she was dizzy, then moved to the edge of the hall and swiftly out the back door. They faced each other on the dried-up grass. He pushed her up against the picket fence and breathed into her ear. His erection pulsated.

'Stop it!' said Arana.

'You could love me. I'm dying inside. I'd give my life for you,' he whispered and tried to kiss her lips. He slid his hand along her breast.

'No!'

A silhouette of a man against the lit doorway.

'Arana!' yelled Steve.

'Get away from me,' said Arana as she disentangled herself from Johnson's arms. She brushed her clothes and walked with dignity up the steps to join Steve.

'What's going on?' he said.

'Nothing, absolutely nothing,' said Arana.

Steve took her arm, and they walked inside the hall.

A bit later, Johnson sensed someone staring at him from across the room. There, over the shredded steamers and rolling balloons in red, black and yellow, was June. He looked back and gave a lame wave, and she laughed and shook her head. With one last lingering sigh at Arana, he skulked out a side door.

On his way home, he thought back to the ball and the handsome dark faces, the size eleven shoes clustered around Arana. Why would she be interested in him? Skinny and white with an overeager grin. He made up his mind: simple, he could seduce her in his room, with incense burning, and a hankie over the bed lamp. Woo her. Force her, even – she would love it. The ripping sound of her panties and the soft dark flesh in his hands, a feeling of marshmallow. She might whimper or beg. He giggled, a puppy sound. All that porn was not wrong. She could be his Black magic woman.

* * *

Steve danced with Arana, and after a while he took her in his arms and whispered: 'You coming with me tonight?'

'You're a bad man, Steve Allunga.' She laughed.

June overheard them, and sniffed, and she wanted to say that Arana was too young, but of course she was not.

'I can show you a good time,' Steve said to Arana.

'Gee, you're a fast worker.'

'You have a nice figure in that gown, true. Be great to see you without it,' said Steve.

'I bet.'

'You know, one day I want what the whitefellas have: a good SUV, you know, a hybrid electric model, and a good comfortable income to have a family. Security, maybe a cruise overseas. I am moving up in the world.'

'I thought political action for the river was your thing?' she said.

'Not tonight.' He smiled.

'I've decided that I am going to chain myself to the water pumps at Big Cotton and get NITV to film me in a hunger strike,' said Arana.

'I'd support you. But tonight, I got other ideas. Coming?' said Steve.

'What, now?'

'Would your mother mind if I took you home tonight?' said Steve, and he didn't seem to see William shaking his head at him.

'Nah. I'll go and tell her.'

They approached Merle. She drank soda water and nibbled peanuts.

'Mum, I'm going with Steve,' said Arana.

'Don't be too late. Twelve o'clock,' said June.

'Make it a bit later?' said Arana with a pleading smile.

'All right,' said Merle.

June put her finger up and moved it back and forth. 'Make sure he gets off at Redfern,' said June under her breath.

The women watched their girl leave the ballroom and sighed.

June reached for her glass and caught William's eyes. 'You haven't misappropriated all that river money, have you, Will?' said June.

'Who told you that?' said Merle.

'Look, sis, won't lie to you. I'm ashamed, but I've put it all back. Believe me, it won't happen again. That horse was a sure thing. Old Blackett gave me a loan with twenty per cent interest. He's not such a bad bloke,' said William. 'I'm paying it back in instalments. I know, I got a problem.'

* * *

It was late and quiet on the outskirts of Wilga when Steve and Arana walked on the riverbank where the levee bank was high and there was the peaceful murmur of night birds. The huge old gum trees, river red gums, leant towards the bank. Those trees were listening and bearing witness. Millions of stars in a dark blue sky.

'Lot of stars out tonight,' he said.

'You going to kiss me or what?' Arana pushed her head up to his and nuzzled him.

'You're playing with fire, woman.'

'I hope so.'

They kissed and held each other while leaning against a tree.

'You're dangerous, Steve.'

'Eh, it's only me. You're safe. I will never hurt you.'

'Something is pulling me to you, in my sleep ... all the time ... I'm going gwarnghi, crazy,' she sighed.

'You reckon I've caught you? Sung you like the old fellas?'

'Maybe ... You make me shiver. Hold me,' she sighed.

Steve took off his jacket and shirt and Arana kissed his chest.

'You gotta be fearless,' he said.

'I can be – when I'm with you.'

Steve unbuttoned her dress and it slipped to the ground.

'My heart's beating so fast,' she whispered.

'I want to take you up north to my people. Sleep under the pandanus,' he said as he laid down his jacket and pulled her gently down beside him. His hands traced her skin and breasts. They panted and sighed and laughed.

'Eh, Arana, this is the first time I've felt like this about a woman.'

The young couple made quick urgent love on the gum leaves; they were breathing hard and laughing then they lay together with arms around each other. They made love again.

The sky was full of rolling stars of the Milky Way, and they pointed out shapes in the dark parts of the Dreaming snake.

'We saw Madukka once by the river, in the moonlight. A huge snake all curled up. Full of magic and scary,' whispered Arana.

CHAPTER 30

21st August

Late at night the cold streets were bare, with just a few dogs walking along the road. The shops were locked up behind steel mesh, and it was as if everyone was asleep or dead. Under a streetlight Steve and Arana kissed while waiting for a taxi.

Steve put his jacket around her. 'I'll get you a taxi,' he said.

'You'll be lucky, there are only two in town,' she laughed.

'I've got the Uber app.'

'No worries.'

'You're beautiful.'

'So are you,' she giggled. 'You nearly fell in the riverbed. Should've seen your face.'

Steve pulled her to him, and they kissed.

'Where's that taxi?' Then he spotted a vehicle. 'Here's one. I'll pull him up. Oh, he drove straight past. Now he's turning around.

Look, here it comes. Hey, taxi!' he called out and stepped onto the road.

'It didn't stop because we're Black,' she said. 'Typical. It's the only time you wish you were a gubbah.'

'Wait a minute.' He whistled at another approaching car.

'It doesn't look like one to me,' she said.

'It's too slow for a taxi,' he said. 'Argh, gunjies! They're stopping, look out, here's trouble.'

Arana yelled out angrily, 'What is this, the old South Africa? Go and harass someone else!'

* * *

Blackett and Johnson drove up and looked out the window.

The sergeant yawned. 'Hey, did you hear about the bloke who goes up to a Black girl on a bus stop, sitting next to three tampons? He says, "Who are your mates?" She says, "Don't talk to them, they're all stuck-up cunts."'

'I don't use that word.'

'Probably never felt one either.'

They looked at Arana on the street.

'Shall we have a look? She sounds drunk,' said Johnson.

Blackett shook his head. 'No.'

But Johnson was keen to find something to do, and it was her. He couldn't help himself. 'I reckon we should. Make sure they're on their way. And that she's okay.'

Just then, Bam Bam and Cowboy roared up on their motorcycles and stood near Blackett.

'Evening, Sergeant,' Bam Bam said and mock-saluted his black helmet.

'These young blokes don't understand what we're dealing with. Arrest one Black, you gotta arrest a hundred of the mongrels,' said Blackett.

'Can we watch? Better than Netflix.' Cowboy laughed, and the bikies stared.

'Look, Johnson, I don't need some Black bastard spewing up on the back seat, all right?' said Blackett. 'Go on, find out for yourself. Well, get goin'! Bugger ya! You too, Bam Bam, piss off!'

The bikies took off with a roar of burning rubber.

Johnson emerged from the car, pulling out a walkie-talkie and turning it on.

'Three months out of the bloody academy and you know it all,' Blackett yelled out the window. 'Turning out bloody social workers!' He spoke into the walkie-talkie. 'Thank you for tuning in, zero tango, we have a ...'

Johnson approached Steve and Arana. 'Hey, wait on ...' The constable clipped away his radio and smiled. 'Good evening, what are you doing out this late?' he said.

'Waiting for a taxi. You cops are always hassling us,' said Arana with rising aggression. She pushed herself forward to face the young constable.

'Settle down, I'm asking a simple question.'

Steve took hold of Arana's arm and began to pull her down the street, 'We're okay,' he yelled over her.

Arana's head twisted back. She could not control the anger, and indignation was rising. 'We haven't done anything. Go and arrest some real crooks,' she shouted.

'A few questions.'

'It's apartheid.'

'Leave her alone, please!' said Steve as he pulled her away.

'Sir, do you live in Wilga?'

'None of your business,' said Arana. 'Anyway, you're gammin Wayne, you know me – Aunty has the investigation service.'

Arana stood tall, hands on her hips. Steve stepped behind her as Johnson looked over at Blackett, who was observing from the police vehicle.

'It's my business when you're out on the street this late,' said Johnson, then he whispered, 'Go along with it, please, Arana – my boss is watching.'

'We're okay, Constable, I'm taking her home now,' said Steve.

Steve tried to pull Arana away, but she refused to budge, and Johnson grew a little irritated.

'What's your address, sir?' he asked.

'He lives in Redfern,' said Arana.

'Redfern, huh?'

'So what?' She sighed.

'What's your address in Wilga?' said Johnson.

'You'd remember. You brought your old dog around.'

Steve moved beside his girlfriend, put his arm around her and whispered, 'Cool it, Arana ...' He pressed his fingers around hers, but she pushed him away, and was defiant; her voice quivered.

'I'm sick of it, you can't push us around. I don't care if you know my dad, it makes no difference, it's police harassment,' said Arana.

'You're making this very difficult, miss ... Anyway, my dog died. Probably poisoned with Ratsak.'

'Sorry, Wayne.'

Johnson spoke into his walkie-talkie. 'What's that, Sarge?'

They heard a fuzzy sound from the radio and it squealed, and they all hunched their shoulders at the high frequency.

'We get enough of this in the city!' she spat.

'Calm down a minute,' said Johnson as he took out his notepad and pen.

'You've got to be kidding!'

Blackett climbed lazily from the vehicle and stood next to Johnson. He just wanted to finish the shift with as little annoyance as possible. 'What's going on here?' he asked as he looked the youngsters up and down and sighed. He stared at Steve. 'What's your name?'

Steve tried to walk away. He knew trouble could happen and pulled Arana with him, but Blackett lost his temper. 'Hey, I'm talking to you, Black boy. What's your name?'

'Steven.'

Blackett grabbed Steve by the shirt and wrenched it, so it tore. Arana slapped the sergeant's hand away, and he glared at her with a little smile.

'Steven what?' said Johnson.

'Steven Allunga. Why don't you nick off?'

'Is this because I rejected you? It's ridiculous!' sighed Arana.

Johnson snarled at Steve.

'Righto,' said Blackett, 'you're under arrest for offensive language. Put the cuffs on him, Constable.'

'You can't arrest me!'

'What's he done?' Arana attempted to pull Steve away, but they held him tight.

'Cuff him!'

'But Sarge,' said Johnson.

'You've got nothing on me, you shit!' Steve said.

Too late, the moment of escape was lost. Steve deflected Johnson's attempt to handcuff him and there was a scuffle and Arana jumped right into the middle of it, shoving at the men.

'What's the charge?' shouted Steve. He moved himself to block Arana from the attention.

'Resisting arrest. Fucking offensive language,' said Blackett.

Steve felt a surge of rage and pushed Johnson away.

'And assaulting a police officer.'

'You've got to be kidding,' said Steve as Johnson reached to his belt and applied the handcuffs.

'Get these things off me! They're hurting me.'

'Let him go, Wayne!' screamed Arana.

Johnson grabbed her, and her dress slid down her shoulder.

At that moment, Bam Bam and Cowboy roared back around the corner and stopped their bikes in the middle of the road. They lit cigarettes and blew smoke rings.

'Need any help, Occifer?' said Bam Bam, and he slapped his leg with hilarity.

'I can film it for youse?' giggled Cowboy.

'Nah, but give us a smoke,' Blackett said, then took the whole cigarette packet.

'That's forty bucks.' Bam Bam gritted his teeth.

'Come on, off you go, girly!' said Johnson as he took Arana's arm and shook his head at Bam Bam. He shoved Arana down the road.

'Climb on the back of my bike, I'll take you anywhere you want, sweetheart,' Bam Bam said, riding alongside Arana.

'Leave her alone! Go home,' shouted Steve. He motioned to her to run.

'Get away, Bam Bam,' Blackett said.

'Just trying to help!' And he roared off, laughing with Cowboy on his heels.

'He's done nothing!' Arana shouted. 'Wayne, you idiot, stop it!' She beseeched Johnson, her pink lips pursed, her face distorted in fury.

'Make yourself scarce. Move!' shouted Blackett.

Arana stood and watched Steve being taken away. It was slow motion; she couldn't breathe. He resisted non-violently, stiffening his body as the police forced his head down to push him into the wagon. Steve was manhandled into the back and Johnson closed the door. The sound of banging rang out in the quiet street. Dogs began howling.

Steve cried out to her through the window. 'I'll be all right. Don't worry!'

Johnson sat in the vehicle next to Blackett, who shook his head in disgust, and they drove off, leaving Arana in a cloud of grit and dust.

* * *

'Surname Allunga, eh?' said Blackett. 'Funny bloody name! Well, well, we got one of them. Nice job, Constable. We got ourselves one of those troublemakers. TRG'll be pleased.'

'This is bullshit!' yelled Steve from the back. 'It's ridiculous … You got nothing on me.'

'Just shut up, you! Lighten up, we're here to look after you.' Blackett lit a cigarette. He turned the plain packaging over in his hand. 'My favourite, the rotting gangrene toes. Nice one.'

***

Hours later, a mist rose from the road outside and it was quiet on the street except for night bird cries. An overhead light illuminated a circle of space in the police station. Steve sat on a wooden chair, stripped down to his white underpants and socks with his hands cuffed behind him. 'Give me back my clothes,' he called to Blackett and Johnson, who were writing at their desks. Sweat dripped from his face, droplets hitting the dust at his feet. He tried to recall which police station in Queensland he had last been in. There had been a few, and he was law-abiding.

'Got everything, Constable?' said Blackett. 'His belt, watch?'

'Yes, sir.'

Blackett wrote a report at his desk while Johnson put Steve's possessions in a plastic bag. The sergeant's shape made shadows on the wall before the desk lamp. They looked over their shoulders as Steve stamped on the floor.

'Let me ring the legal service,' Steve said. 'I know my rights. I want to ring the liaison officer.'

'For Christ's sake, it's two o'clock in the morning. Where do you think you are? The big smoke?' said Blackett. The sergeant closed his eyes. He didn't want to be there. He imagined his bed and Mavis curled around her cat, Tuppence.

'Look, be reasonable, just one call. It's my legal right! Hand me my mobile,' said Steve.

'This is procedure, Mr Allunga. You'll have to be patient,' said Johnson.

'Now, what have we got?' said Blackett. 'One leather wallet, one fountain pen, five dollars and sixty cents. A nice new iPhone. What else?'

'Let me go, for God's sake! Let me have my clothes back, I didn't do anything. We don't have to put up with this treatment. What happened to Thommo? Tell us the truth,' shouted Steve.

'You're looking for a hiding ... What else, Johnson?' Blackett thought that at least the young officer might get a lesson in how not to arrest someone for nothing much. He'd have to learn by doing.

'A set of keys ... Hey, mate, keep it down! A handkerchief, ironed, ha. Cufflinks.'

'The law states that I'm entitled to a solicitor. You have to let me call one,' yelled Steve.

'Isn't it procedure to give him one call, Sarge?' said Johnson as he hunched his shoulders miserably and flicked through the police manual.

'Did you get his shoelaces?' Blackett said.

'Yeah, here.' He handed Blackett the laces.

'You taking any medication? No? Right,' said Blackett, and he pointed with his thumb at the space on the form to fill in.

'People are gunna hear about this! You're gunna regret it! Obeying orders, is it, fellas? Covering up a murder. You're working with bikies and drug dealers, like your mate Bam Bam? You didn't even arrest him for that assault!'

'I suggest that you keep it down, Mr Allunga,' said Johnson.

'Keep the black shits down?' said Steve.

'Language!' said Johnson. 'Everyone should do a course in how to use appropriate cultural language.'

'What's this? A bloody silk bow tie! You got a receipt for this?' said Blackett.

'You can't get away with it,' said Steve. 'You think Murries are gunna lay down and die, let you shit all over us? Let you murder us like you did to that fella who lived on the riverbank?'

'I said: shut up!'

Blackett tidied up his desk and laid his balding head on the wood. He almost felt like snoozing.

'Times have changed,' Steve said. 'It's not 1970 now, you white pigs!'

'Take him out back?' asked Johnson, but Blackett looked at him with exasperation. Didn't the lad know anything? Did he have to wipe his backside as well?

Steve wouldn't stop. 'Everyone knows about your little drug ring, you corrupt cops. Thommo never knocked your plantation off – he wanted to save the rivers ... You corrupt bastards!'

'Okay, lock him up,' said Blackett, 'before I bloody teach him a lesson he never forgets.'

'You're all killers. You've got nothing on me!'

'Come on, you, settle down,' said Johnson as he hauled Steve upright and led him towards the back of the office.

'You can't do this,' Steve said.

Johnson pushed Steve towards the cell, but the man resisted violently and dug his hands into the doorjamb.

The young Murri's eyes were wild with fear. 'I'm not going in there. Let me go!'

'Shut up! Get him in,' said Blackett as the struggle escalated between the younger men.

'Come on, mate, don't struggle,' said Johnson.

'Let me go. I'll fight you every inch of the way on this! Don't put me in a cell. Please! I want to call the liaison officer. I know my rights!' shouted Steve.

Blackett stood up with a tired groan and moved to Johnson's side.

'Sarge, I can't get a hold of him,' said Johnson.

'Blackfellas get real slippery when they're stirred up,' Blackett said. 'Come on, put him in the cell!'

Steve yelled and kicked at the police as they grabbed him from behind. He pushed both feet up against the wall and threw Johnson heavily onto his back on the floor. Steve got up as Johnson rose slowly and punched Steve hard in the face. Steve's fist went up but Blackett pushed Johnson out of the way. Johnson fell again and twisted his arm and hurt his hand, badly.

'Can't you do anything right, Johnson?' yelled Blackett. He turned to Steve. 'I said get in the cell!'

Johnson was on his knees, humiliated. He got up and his hand felt broken; the pain was severe. He watched with a gaping mouth as Blackett picked up Steve then let him fall to the concrete floor.

'You little shit!' shouted Blackett.

Johnson walked out of the cell and lay his miserable head on his desk.

Blackett watched Johnson leave and then raised his fist, and punched Steve repeatedly in the face, a heavy boxer punch designed to hurt. Vermilion drops of blood trickled down Steve's forehead. Steve flopped heavily onto the ground with a thud like a cracking rock, and a wound opened in his head. Blackett pulled Steve up again and marched the groggy form towards the cell, but Steve was almost unconscious and moaning, and blood gushed, making a dribbling red puddle along the office floor.

Johnson stood in frozen shock. 'It wasn't meant to be like this. I wanted to help them. I should never have joined the force. Oh God, I'm sorry,' said Johnson.

They heaved Steve onto the concrete floor of the cell and closed the door. Dark blood was pooling around his skull.

* * *

A half-hour passed. Johnson had a bucket and mop and swished water around the office floor as Blackett sat at his desk struggling with a computer, cursing the machine.

Johnson had looked in on Steve through the peephole, but he seemed to be asleep, so Johnson sat back at his desk and wrote out some notes.

Later he took the keys to the cell and opened the door. He saw Steve still on the floor, and he leant down to take the pulse. It was very faint, and Johnson's face lit up with panic. 'Not responsive! Prisoner not responsive! CPR, we gotta do CPR or something!' said Johnson.

Steven Allunga was dying softly with warm liquid around his head as the police scrambled to revive him with useless puffing into his bluish mouth. Blackett pumped the chest like he was trying to blow up an air mattress with a cheap two-dollar-shop hand pump. Did Steve die with a gunjie's mouth on his? Lip to lip. Saliva joined. Johnson breathed in. He turned Steve's face to the side like he had learnt in training. The earthly life draining away. The quiet broken by husky voices, and Johnson stopped the mouth-to-mouth and looked at his boss. A still beauty in the young dead man's face. A serenity, almost a peace.

'Twenty-eight, twenty-nine, thirty, you take over, Sarge,' yelled Johnson. 'Where's the resuscitator thingy?'

'He's gone, mate. Poor bastard,' said Blackett.

Blackett was oddly quiet now, and he reached into his back pocket for his cigarettes, but first he took out the ironed handkerchief that Mavis had placed there to wipe the gore from his hand. Mavis did the washing and ironing, and she liked a neat pair of trousers.

The hooting voice of a distant owl. Always a bloody owl.

The stench of blood, like an abattoir. That aroma got up your nose and you couldn't get rid of it.

Now the two police gazed at each other on their knees, like lovers in the terrible intimacy.

No sign of life from Steve.

'Time of death: 2.30 a.m.,' said Blackett as he checked his watch.

Their lives were unravelling before them.

'Shithead,' muttered Blackett.

'Me?' said Johnson.

'No, the dead shithead.'

'Don't disrespect the dead, Sarge!' He turned his head away and tears trickled. He blew his nose and blubbered.

'Oh, now you want to be all respectful, you idiot. In case you haven't noticed, you just killed him.'

'It was …'

'Yeah, an accident. You're a fucken accident!' snarled Blackett.

The smell of dust and mildew. The clock ticked and ticked, and they just looked at the corpse.

Blackett sprang into action, cursing his painful knees. He ordered Johnson to get more water in the bucket and to rinse the mop. 'And call the medics,' said Blackett.

'Who?'

'The ambos, you moron. The number is on the wall. And for Christ's sake, hurry!'

A hot dust-filled breeze poured through the cell bars. Blackett thought he might pray. But to what or who? Just to get a quick message to God; it might help. It was a bad occurrence; it was going to sound terrible in front of the magistrate. He might lose his pension. His teeth clenched the cigarette. He was terribly sorry. The smoke eased him. He watched Johnson wringing out the mop and the water going pink. There was so much blood. Blood drying on his knuckles. He scratched the cuticles, the mess congealing between his fingers. He grabbed the mop from Johnson and kept mopping in circles.

Johnson stood shellshocked by the desk while Blackett finished writing the report. The sound of the keyboard going *clackety-clack*. Mystery and fear of the unknown.

'This town's a dead-end shithole, time to get out,' said Blackett. He used to imagine himself hand in hand with his wife on the beach at Woy Woy. He would have got a boat and fished for bream. The town of Wilga was too hot in summer. It always felt like it would explode in a fiery furnace. Bushfires. Or melt like liquid fat, leaving a nasty, greasy stain where the people had been. Outside now, woof woof woof, a dog barked relentlessly in the street. It was waking everyone.

Like a pain in the arse, the computer had frozen, and Blackett thought, it's all shit, that his life had turned to shit. There were red spatters on his boots, and he bent down to wipe them off with his thumb. He was nervous and his face contorted with the pain in his chest. He took some painkillers and wiped up any stains from around his desk with piles of tissues. He sharpened a pencil.

Still, the corpse of Steve lay uncovered and aching for some sign of compassion. The pale feet were sticking up naked, like Jesus' feet on the cross.

Madukka, Waaway, the great spirit serpent already approaching to take him home. It had many names and manifestations. A glistening giant rainbow snake that could eat people and create life and disgorge its children and all animal life. It churned up the soil, left sighing fissures where creeks and billabongs and mighty rivers began. The eye of the serpent blinked at Steve at the moment he drew his last breath, luring him away, soothing and caring, because no human could, at that moment of death, offer him succour.

Johnson paced up and down. Blackett watched him; being a trained copper helped him with resilience. Johnson held on, biting his bottom lip. Tasting its drop of blood. He thought later he might be blubbering on his bed, the surrounding sheet sopping wet.

'My mum will be disappointed,' said Johnson.

A sound from the main office alerted Blackett to someone waiting at the desk. He wiped his face and went out to the charge desk. Bam Bam and Cowboy stood smiling in the bright light.

'What do you want?' said Blackett.

'Nothing. Just passing by. Thought you might want us to pick up a kebab or something?'

'Not now, mate, we are very busy with a serious incident.'

Blackett stepped out from the desk and opened the station door. 'Out!' he said.

'We have a right to be victims too, you know,' said Cowboy. 'We pay rates and stuff. Well, he does. I'm on a disability pension.'

Bam Bam stared at the mess on the floor, at the mop and bucket full of something red. 'What's been happening, Sarge?'

'Nothing. Get the fuck out, now!'

CHAPTER 31

22nd August

Arana sat on the street and worried for a few hours, not knowing what to do. Then she walked back to June's house. Her hair stood on end and her face creased in anguish as she banged on the door. It was very late and everyone was asleep. The pallid grey sky filled with dark tumbling clouds, and that snowy owl sat immoveable above the lintel.

Inside, June lay in bed tossing in the oppressive atmosphere. She heard crying, and as her eyes adjusted to the darkness in her room, all her fears had become real. The thumping on the door, the sound of something terrible, a call of pain.

'Mum! Mum!' Arana yelled out as Merle went to the door, half-asleep.

'That you, Rara?' said Merle.

The door opened, and the girl was a banshee with staring eyes. 'Mum, you gotta come down the police station, they've got Steve!'

'Oh no, what's he done now?'

'Nothing!' shouted Arana.

William walked out in his flannelette pyjamas. 'What's the big noise? You're waking up the whole house,' he said.

'I'm scared. Please come get Steve out of gaol!'

'Ah, someone always gets arrested after a ball,' said William.

'It's late, love, we'll go in the morning,' said Merle, but June came from her room and could see that Arana was frightened and she wanted them to leave for the police station right away.

'Okay, darling, let's go. I won't sleep anyway,' said June.

'You don't understand, they've got it in for him. We know something about Thommo's death. Please help him!'

'That Steve is trouble, bloody flash mouth, shouldn't do investigations, leave to police,' said William.

Arana began to cry hysterically. 'I know he needs us; I can feel something. I'm begging you to help him.'

'We'll ring up. You go to bed, it's not your worry,' said William, but June grabbed her purse and hurried into a pair of jeans.

'We'd better go now,' she said.

'All right …' William said. 'But if we go down there and cause a stink, he might cop a hiding … Come on, my baby, he'll be right.' He patted Arana on her back.

They drove off in William's car, and the sound of a mopoke bird pierced the night. They all looked at each other. They hated hearing that bird. It scared them, and the fear rose inside the car. Merle had June by the hand and was crushing her fingers.

* * *

At the police station, Blackett gawped at Johnson as he sat hunched in a chair.

'Are you praying, son? What for? Guilty feelings? You might ask for forgiveness. That goes down well. Or are you just clenching your fists because they brought you up by a Methodist mother and you remember performing the rites like a trained seal?'

'Helps. Maybe you should give it a go?' said Johnson.

'No, I'm a closet atheist. Wife doesn't know.'

'A pity you didn't have kids, Sarge.'

'Why's that?'

'You might have a bit more compassion.'

'Stick your compassion, that's for weak people.'

Blackett handed Johnson a broom. 'When you finish clearing up this office, you go get me something from the trophy room. A long one,' he said.

'No, Sarge.'

'You disobeying an instruction, Constable?'

'But the ambulance is coming,' said Johnson.

They heard the siren wail like a baby.

'Get a move on.'

Johnson left the room, and came back with a football sock. Blackett snatched it. They both entered the cell, quiet as death and sticky with pain and blood. Johnson nearly vomited, and he bent forward with his eyes covered by his hands and a choking sob burst out.

'One thing, sweetheart,' said Blackett. 'I'm not losing my position because of the stupidity of a snotty-nosed kid like you!'

'I can't . . .'

They stood motionless over Steve's corpse, the iridescent light in the room weaving around their heads. The night was suddenly freezing cold, and Johnson sensed icy hands at his neck. His hair stood up in fright and there was no end to how sorry he felt.

'I'll tie it around his neck,' said Blackett, and he leant down in a matter-of-fact manner to tie the sock. Johnson could not watch, and he let his head fall forward.

'It'll appear like he was in a fight,' said Blackett, 'then he was in the cell and . . . he took off his sock, oh yeah, put the other one on his other foot, move!'

Johnson pulled the sock over the other pale foot. He tried to not see his own hands pulling at the immoveable toes. The young constable had flown out the window in his mind and was tearing across the paddocks back to Newcastle and into his mother's arms.

'Yeah, he fell from the bars because of his weight,' said Blackett, 'and he hits his head on the floor – that explains all the blood in here.'

'Yes.'

'That's right. Do as you're told. We can fix it. Don't go all gooey on me. Be brave, for fuck's sake.'

Blackett tied the noose while Johnson placed the other sock on the pale foot. He put Steve's body against the bars and his head tilted like a doll. The police left the cell and Blackett washed his hands in the bucket and dried them carefully on his handkerchief, which he then folded and pushed in his back pocket. He studied Johnson as he stood limp and impotent in the doorway.

Without warning, Blackett pulled back his arm and landed a heavy fist in Johnson's eye. The young man reeled backwards and stumbled.

'Shit! What's that for?' said Johnson.

'Righto, now you look like you were assaulted. You can write that up,' said Blackett as he shook his palm. Then his fist clenched again.

Johnson held his face in fear. 'Don't hit me again, for Christ's sake!' His hands came up in supplication.

'A few bruises won't damage that pretty mug, mate. Now, get out of my sight and let me write the report. The senior sergeant will be here soon – we're only two hundred kays from his town. Go on, get out! And don't wash that cut – it's gotta look like it was a fair fight.'

'You shouldn't have hit him so hard,' said Johnson.

'Don't know our own strength, do we? Just go, start writing in your notebook. He hit you, okay?'

Johnson hung his head and sighed. 'Doesn't seem right. Not the truth,' he said.

'Who are you? Jesus? You want me to bring the Gobblers in to make it look real?' Blackett looked down and began to type. 'We need to get this right. Two a.m., a fight. We put him in the cell and you checked on him at two fifteen a.m. Then again at two thirty a.m. and found him dead. Okay? Remember those details exactly.'

* * *

On the road from June's house the family heard sirens from somewhere, and there seemed to be a crowd of Murries running towards the centre of town. June looked at the poor caged emu in the park and shuddered. She hated anything in a cage.

The streets filled with blackfellas. Mostly young men, carrying sticks and bottles.

The sky turned a strange mauve colour, still full of black clouds. It might have rained; a cold wind blew with thunder. Biamee the sky god rumbled. He might be out and about. A big gust of wind hit their faces as leaves and dust tossed in their eyes.

William, Merle, Arana and June drove through the streets towards the town centre; they parked and stopped to breathe. June looked over at William and his face was purple. The colour rising and he couldn't breathe.

'God, William don't have a bloody heart attack now, we haven't got the time,' said June.

He laughed and coughed.

'Hurry! Come on, Dad!' Arana helped him keep moving.

'None of this would have happened if you'd come home with us,' said Merle.

'They'll just give him a thumping and let him out tomorrow, nothing to worry about, he will be all right,' June said, but she didn't believe it for a moment because a terrible feeling had them all by the throat. Knowing something unthinkable had happened.

The family watched the main street and there in front of them was a white vehicle.

Arana stopped dead in her tracks. The ambulance stood outside the police station, back doors gaped open.

'Dad . . . An ambulance. Look!'

'Oh, my God! Trouble, all right,' William said and stared at June.

'We've gotta help him,' said Arana.

They rushed into the station, where Blackett sat at his desk, shuffling paper. Like it was a normal night. Everyone froze and stood in front of the desk. Papers were strewn all over, and the sergeant sat, a peculiar light in his face, as though he had been electrocuted. His grey hair hung in tendrils over his forehead, and he sweated big drops onto the desk. Plop, plop. All quiet in the room, except for the ticking clock. He looked up.

'Yeah, what can we do for you?' said Blackett.

'What have you done with him?' said William.

'Stand in an orderly fashion back from the desk.'

'Allunga.'

'Are you family?' He watched them, but his head tilted forward as though weighed down. His hand shook, and he tapped a packet of cigarettes.

'Not exactly. Friends,' said June as she stepped forward to face the police officer.

'Well, I'm afraid we can't tell you anything. We're busy, sorry. Do you know where his family live? We would appreciate knowing who are the next of kin.'

'Yes, I can find his family's address. Where is he? Is he all right? I want to see him, please, sir,' said Arana.

Blackett didn't budge and went back to writing. His hand twitched.

Arana flipped open the wooden barrier and pushed past Blackett towards the cell area, just as Johnson emerged with a terrified face. The young constable wore a ripped, blood-stained shirt, and he ran his fingers through his hair. His face told them everything.

Blackett reacted by standing up and getting between Johnson and Arana. He shooed his constable back towards the cell to block the way, as if ushering a dog into a cage.

'No, you can't go in there,' choked Johnson. 'William, what're you doing here?'

'What's going on?' said William slowly and calmly.

Arana darted out of Johnson's way and pushed towards the back-room door, her arms flailing in the air.

'Stop her, William!' said Johnson. 'I'm sorry, the paramedics are in there. She can't help! No one can help.' He gently grabbed Arana and pushed her back behind the front desk and closed the barricade. She stared at it.

'Jesus, what's goin' on?' William said. 'Wayne, what's your mother gunna say?'

'You'll have to speak to the sergeant, he's in charge,' said Johnson as he swallowed and assumed a professional face. He avoided looking at the family. Then his hand covered his face. It seemed like he might cry.

'Look, we want to make some inquiries,' said June with authority. 'You've got Steven Allunga, right? You took him in last night. We want to see him right now.'

'Why is he here?' shouted Arana. 'He's done nothing wrong. We were just waiting for a taxi. Mum, Dad, do something!'

Blackett ignored them and calmly resumed his paperwork. 'As you can tell, we're very busy . . .'

'Why do you have to contact his next of kin?' said Arana.

'He's not dead, is he?' June whispered, and the words hung in the air.

Right on cue, an owl hooted through the open window. They all stared at each other, their faces in horror.

Clear now, a spirit spun into the room. It must have been an alarmed, frightened spirit; it couldn't rest; it paced and cried. A fresh, bloody and sad spirit that made the humans shudder.

'It's none of your business,' said Blackett. 'I'm sorry, but we have to follow certain procedures. You can call up tomorrow and we might be able to help you ...'

'We'll just sit here and wait,' said William, and he sat and he patted a seat for Arana.

She sobbed and sat down with her head in her hands. Merle leant against her, sighing.

***

The family had been waiting in the dingy building for two hours with no information. Blackett and Johnson were at the desk just near the waiting room.

Arana had collapsed in William's lap. June looked at the torn fabric and bursting rubber on the seats and wondered how many poor, wretched Murries had sat here waiting in vain for some justice. Time crawled along, and the sound of crickets outside punctuated the quiet. Off in the distance they heard someone crying, but they didn't know who.

Inside the police station interview room, the two officers filled out the paperwork.

'Just do as you're told,' said Blackett. 'Read it for me, Constable. Slowly.'

June observed Johnson through the doorway as he read with perspiration streaming. She couldn't hear what they were saying. He showed obvious exhaustion. Blackett smoked and wrung his hands, back and forth.

'The deceased,' Johnson said, 'one Steven Allunga, had shown signs of intoxication. He had been shouting abusively and had repeatedly thrown himself against the door and walls of the cell, displaying an inclination to inflict self-injury. He behaved in an uncontrollable manner, then became quiet after a while. When I called in for a routine inspection, I noticed the deceased. I yelled out for assistance. The sergeant got the keys from the charge room desk and—'

'No, you got the keys. Write that,' said Blackett.

Johnson kept reading. 'I got the keys from the charge room desk, and we opened the cell.'

'Good, yeah, go on.'

'Upon entering the cell, we became aware that the deceased had suspended himself, and that no other person was directly involved.'

'Yes, yes, that's right,' said Blackett.

June stood up and placed herself by the open door. A crack of light. She knew that someone had to witness whatever those men cooked up in their office. She listened at the door and William kept an eye out for anyone coming. She stood still as stone and her eyes glowed.

Johnson continued. 'I held the deceased up from the wall and I untied the sock from which he was suspended. We laid the deceased on the bunk. The Sergeant undid the sock attached to the neck.'

Merle held Arana as she slumped in her arms. She seemed to be unable to breathe. Her mother loosened her dress, and the girl began sobbing.

June and William listened at the crack in the door.

'The sergeant put his hand on the deceased person's body and found that he felt cold and had no pulse. I attempted to take the pulse at the neck but could not find it. The time was three a.m. I determined it not feasible to continue cardiopulmonary resuscitation.'

'That's good,' said Blackett.

'It's clear that the deceased had an intention to take his own life, knowing the consequences of his act,' said Johnson. He stared into space, then put his head in his hands.

'That'll do. It reads well,' said Blackett.

'It's not right! Sarge, I don't want any part of it.'

Blackett hissed at the constable. 'What don't you want any part of, Constable? Your career, your future? Want to throw it all away? If they start digging around, they might even charge me for young Thommo's death. Conspiracy or something.'

'I don't know ...' said Johnson as he leant his face on the desk.

'You were doing your job, son. We used no more force than necessary on the deceased ... Come on, I did this for you. We got to stick together on this; they'll never damn believe us ... It's the way the force operates. Now, go and give the medics a helping hand before I lose my temper.' Blackett looked up to see, the family outside watching him in shock.

***

William and June looked at each other, mouths open, but nothing came out.

June snapped from her spot and pushed the door wide open. 'Deceased?! Who is deceased? What have you done?!' June cried out.

The police officers gazed in surprise at her, then pushed her back and the door slammed shut again.

June waited a moment and slowly twisted the handle again so those inside could hear her trying the door.

'My God!' June whispered. As if a death could be nothing – routine stuff, destroying a young life. His ghost seemed there with them, fresh killed, hovering above their heads, murmuring.

'You killed him!' screamed Arana. 'Murderers, murderers!'

June held on to her as she wailed. Then Arana screamed, and shook, and tried to get into that cell. Her fingers scratched down the surface of black metal, the sound a terrible grating in their heads. She rushed back and forth, yelling and hitting the door, and William and June could barely hold her. They wanted to whisk her away and tell her it was all right. But, of course, it would never be. Sudden death, like a flash flood, washed everything away. Arana continued to show huge strength as she trembled in pain. The family wanted to melt pain away. June shook too, and the blood in her head boiled like a volcano. It might tear her skull apart; her brains would burst like lava onto walls and floors. They would have to mop it up, with no recollection of why it had happened.

June remembered she had seen a large flock of white cockatoos descending on the edge of the town, a sign of coming death.

William tried to restrain Arana. He grabbed her flailing arms and hugged her to his chest, just as Blackett walked into the room.

'Look, William,' Blackett said. 'I'm really sorry to have to inform you about an incident. These things occur. A lad has passed away; it's a tragedy. Terrible. But we are doing all we can. Just go home, let us get on with our job.'

'Steve!' said Arana. 'Mum ...'

Blackett pushed them all towards the door. 'Constable, show them out, please,' said Blackett to Johnson, who stood with a frozen expression.

'Listen ... please ... go home,' said Johnson. 'William ... you gotta go. I'm sorry, I don't know how it happened, I don't know.'

'What happened? Your family will want to know what happened. You come from a kind mob, Wayne. You're not like him! We need the bloody truth.'

'An ... accident. I don't ... I can't talk about it. Oh God, I'm in trouble, I am.' He straightened up and breathed in. 'You're ordered to leave the premises.' Johnson ushered them out the front door.

Arana tore away and pounded on the locked door to the cells out back.

William quickly grabbed her back and held her tight, all while staring at Johnson. 'Wayne, I'd never believe a quiet kid from Newcastle might end up a bloody killer,' said William.

'You got to believe me. I couldn't ... It wasn't like that. One moment he's kicking, shouting and then ... It happened so quick. Oh, shit!' said Johnson.

'Quiet, Constable!' Blackett shouted. 'Don't say anything, you stupid fool!'

Arana tried to break her mother's hold, and fought to go to the back room. 'I need to see, just look at him. Please, Mum, I've got to look at him!' Arana cried blubbery tears and hung on to William's shoulder.

'It's all right, my baby,' soothed William as she collapsed in a semi-faint. June felt a fierce rush, her heart going like thunder.

'We'll be calling the legal service!' June said, and shook her fist. 'You can't go around murdering Blacks, even if you whitefellas been doing it for two hundred years. We want justice, and by God we will get it!' June suddenly released, and the emotion subsided to a quiet fury.

'We cannot confirm or deny that there has been an incident,' said Blackett.

William moved right up to Blackett's face, his chin almost touching the sergeant's nose. They breathed each other in.

'I'll contact his family,' William said. 'They'll be wanting to talk to you. We'll be back.' He stepped back and his fist rose, but Merle moved his hand and held it down.

'Not now, Will,' she whispered.

'No one should come out dead,' said June.

Merle and William turned to leave, supporting Arana, who had collapsed to near unconsciousness.

June stood staring at the gunjies. She hated them so much. She wanted to be a superwoman and rip down the whole building, flinging their stupid reports and office furniture to hell.

The family left.

'There's gunna be bloody trouble,' said Blackett, and he piled papers in a folder and clipped it down with a loud click. 'Get on the blower.'

'The look on their faces . . .' said Johnson.

'Do your bloody job! I just gave you an order, son!'

Johnson was mute. His face was whiter than usual; the blood seemed to have gone completely from his body. He stood like a ghost.

***

The family drove home. It seemed to them as if nobody cared.

The house stood open, and the door swung in the cool night breeze. They sat in the kitchen, but no one could rest, and after cups of tea they sat long into the night. Not knowing what else to do, they wrote a letter asking all the questions that had been ignored, folded it, put it in an envelope to take to the police.

June asked Merle to keep Arana in the house and watch that she did not go anywhere.

Arana held out her hand to her. 'The last moment I saw Steve, he squeezed my arm and looked into my eyes. The gunjies had him handcuffed. He said, "Don't worry, I'll be back soon." His outline against the flashing blue light. Looked real small up against the police. Last time I saw him alive. I knew something bad was going to happen. I sensed it all night.'

'It's all right, Arana,' June said. 'We'll find out what's happening. Try to stay calm.'

After an anxious time, they could stand it no longer and they went back to the police station with a letter. William, Jack and June sat for hours with their heads down, waiting, waiting. The grey walls dribbled sweat and misery.

'Ballun place now, death place,' Jack shuddered. 'They better get a wirringan Murri doctor to clean the wandabaa outta here. Gagil, no good,' said Jack.

He paced around, smoking a rollie.

* * *

Back in the police station at 6 a.m. Daddy-long-legs spiders ran up and down the dirty wall as the family listened to scuttling sounds. A snake had got into the eaves, after mice. Probably a python. These tiny things focused June's mind.

One of the police officers put his hairy hand on her shoulder. It pushed like lead, and she shuddered. His other hand held a pile of photocopies and a ring binder, but he ignored their letter, despite William continually pointing to it and shoving it towards Blackett.

'This is the ironclad case against him,' said Blackett as he put the folder on the desk and slapped it.

'What? That's two young men dead! Who cares what so-called case you have?' June said, and felt like throttling him.

Jack rose up to his full height and shook his fist in Blackett's face. 'That wandabaa gunna get you! And he'll come for you, a wirringan, and he will point the bone and you will shit yourself!' said Jack. 'You'll just keel over and fuckin' die. They got feather feet and move like lightning.'

June pulled Jack back down to the seat and held his hand. 'It's all right, Unc, calm down, we're doing all we can,' said June.

'You think you'll get away with this? It's murder!' said William, and he stood up against the wooden railing with his hands gripping the barrier.

'You reckon you're smart, don't you?' said Blackett to William as he gazed at June. 'Your wife went to some university? That right? Teacher? And that deceased person was a radical. You teachers, I hate you all. Bloody caning, left, right and centre. Why aren't you lot down in Sydney, instead of making my life difficult?'

'Steven Allunga and Thommo are dead,' said June. 'We're not leaving until we get answers.'

The sergeant walked about the room. He arranged some tins of pencils and wiped his nose with a handkerchief. He opened the grey door to an inner room. He indicated it with a jerk of his head. 'You've seen our trophy room?'

June gazed at the wall inside the room. It had rows of football socks hanging from hooks. She looked away. She hated him and all that he stood for, with his white prejudice, and his cruelty and power over them. He even laughed. She couldn't believe that he was saying those things.

'They neck themselves, these young blackfellas. Don't know why; it's a shame; it's a national phenomenon and we don't have the resources to deal with it.'

'That wandabaa is coming for you!' said Jack as he strode out of the office and into the street. 'Kapili your head. I'm not listening to his bullshit!'

In the office, Blackett's mouth was working at a broken tooth, and June hoped that he might be about to make some kind of statement, but he just stared at them and was deathly quiet. The tongue searched around and around that ugly yellow tooth. In her politeness, June stood perfectly still and stared back at him, with her hands on hips, and it made him shiver. Her searching eyes could turn a man into water. Perhaps he was suffering from some kind of paranoia. Perhaps he would lose his job or be suspended or go to gaol himself.

'You haven't given the family a copy of the report,' said June.

'Not your business. Are you his family? Are you his mother? No, you're not! You lot are not worth a mention in these proceedings. No power, no rights, no nothing. So, piss off! Before I throw you in a cell!'

June repressed a sob in her throat. She went quiet and the big clock on the wall ticked. Darkness brooded in corners, and Steve's body seemed to float in the heavy air. It was like all of them were beginning to rot. Rot like a dead body laid out in the cold on a slab out back of the mortuary. A grey sheet hung overhead with pale skeletal feet sticking out.

He would be beginning to show signs of rigor mortis. Where did the blood go from a dead person's body? Did it somehow evaporate? Or did it puddle deep in the organs? Did he have a bloody face? Why? Was blood vomited up? No, he was supposed to have died from asphyxiation. Why wouldn't they say? Some young people died from heart attacks, but not him, surely, he was so fit. The family should be asked to view his body in the morgue. His family were so far away; it was hard to imagine them travelling anytime soon. Would his mother have enough money for all those flights? What would they find? Something so terrible they might collapse and hold on to the cold mortuary steel bench. Seeing only his shell. Pale and icy, like a chicken in the fridge.

Looking at the corpse through a crack in the door, wondering where the lad had gone. He might have left for footy practice, not be bound up in a black plastic bag heading for a coroner's van like a lump of butcher's meat. June thought about Thommo and the sound of a smash to his head. No proper burial. She pictured Steve's lovely face pushed under plastic, and she wanted to reach through the grey walls and pull that bag down so they could see his face, so he could breathe – but of course he couldn't. Her hand would linger over him and touch the cheek as though to say, 'It's all right, young fella, we're here waiting for you. We will accompany you to the door of light.' But she was no Christian, so exactly what image was that? A door or pearly gate with holy angels and an old white man with a beard and a book? The big archangel Michael with feathered wings from the illustrated children's Bible. With a huge sword to smite . . . What the hell was 'smite'? No, this young Aboriginal man would depart alongside his ancestors, and they'd dance and sing in old language, and the family would cover his grave with shells. It was a comfort.

Then, another bloody siren, enough to make their hair stand on end. William pressed her hand. That siren scream heralded disaster. June pushed her fingers into her ears to cut out surrounding misery.

***

Steve floated above the room. He wanted to stand up, but dizziness won. Searing pain. The office phone rang and he wanted to pick it up. In a sticky bloody fist and say, 'Hello.' In a whisper, asking, 'Who do you want to speak to?' His body was useless. He looked at his arm, reaching but not moving. A store dummy, only black. Unable to make it move. It hung, something not of his body. He would like to fly north home, like a migrating duck.

He was going back to his father's true Country. Near where? Now a lost place. He stared from beneath the grey police cell ceiling, just near a flyblown single light bulb. Cobwebs and grime. He observed Blackett's balding head from above as the man mopped up blood with a cheap Woolworths squeegee mop. The younger copper sat in a pool of blood, whimpering. Poor thing. Steve wanted to reach down and touch his shoulder and murmur, 'It's all right, I understand, you didn't mean it.' Blood was everywhere on the lino, its pattern of paisley, 1970s Carnaby Street. His fatal flaw. What was it? The hubris? The right word?

Steve's spirit zoomed across the room and arched, an astronaut staring out the window. He saw Arana's family walking away and he wanted to call out, 'Wait for me,' but nothing came from his mouth. He watched the cleaning of his body below. The arm lifted and sponged somewhat gently. Suddenly knowing that he had left his flesh behind and had an urgent need to go somewhere. To find her. Comfort her. He waited for the spirit people to lead him into the sky. Half-bird, half-human.

He lay on the floor while the world exploded, and then he travelled upwards, ever upwards. The eyes flickered open, and a monstrous face breathed near to his. Blackett. Steve felt as if he had drowned in blood. The alien face floated. Steve's hand stretched towards it and fire came pouring into his skull, and in a blink, gone. Earlier, he had seen the look in the young copper's eyes. Fierce and terrible, looking older than time. Now, small gifts floated in the mystic air. A coolamon, a firestick, a dillybag full of orange feathers. He had fallen in love with Arana. He flew down the hollow fissure, deeper and deeper to that light. She had given herself to him. Her first love. Her body and spirit. But now he had to leave her in incredible sadness. He looked at the handcuffs on his wrists, down there on the concrete. Blood oozed from a wound in his head. Hair and flesh. The thick steel door kept him from flying to her. But surely he might have drifted through it somehow. Inhaling her sweet-smelling hair and burnished gold skin.

* * *

At last, coming out of her dazed state, June snatched the letter from the desk and folded it into quarters. 'We will take our inquiries higher up!' said June firmly.

'Yeah, where?' said Blackett. 'To God? To Jesus Christ? Or the Pope, perhaps? You do that, ma'am.' Blackett unfolded a piece of paper and read it out like a zombie. 'This has been traumatic for all involved and for the next of kin. These matters will be under investigation by the state coroner, and no further comment can be made at this present time. Any member of the public who thinks a crime has been committed should make a formal complaint to police through the appropriate channels so it can be thoroughly investigated.'

It was enough that the family sat terrified in the waiting room, enough that they had begged for answers. June sensed a rage beginning, one no one would be able to stop. A rage at the white man's treatment of their people, from Major Nunn's attempted genocidal extermination of the Gamilaraay at Waterloo Creek in 1838 right up to now. Blood had poured into that creek and bodies were flung in piles and burnt. Children were run through with bloody sabres that now hung proudly in town museums next to bronze statues of that British major. Aboriginal women had been raped and tortured and raped again for fun, and then murdered in an ecstasy of bloodletting, ensuring that no more of those little Black bastards were born to resist the onslaught of British colonial supremacy. Then the years of slavery and imprisonment. Forget the past, they all said. But enough was enough.

June and the others walked back to the car along the restless streets where young Murri men gathered in groups and called out to each other. The air was thick with misery and fuming anger. June ran her hand along the park's emu cage and saw herself jumping over the fence to let the poor creature free. 'Run, dinewan, run,' she whispered, and leant against the black metal fence.

CHAPTER 32

23rd August

Into the sky, Steve's spirit tumbled in the light spring air. His life rushed through him, a television program on speed. He sniffed smoke and cooking fish and turtle and touched the heat of the northern sun. He saw the flap of the flags on the graves of Yolngu grandfathers. Red and white flying in the sea wind. The wind brought black clouds and rain. He loved that place. He was confused with ceremony, the beating time of clapping stick and yidaki. He wasn't keeping up with the tumult of spirits. He begged to lie down, to rest and not endlessly soar. He lost the sacred dillybag. His initiation spirit home. How could he find the right songs? It hung safely in a pandanus tree on the beach. The delicate orange parrot feathers glistened.

His young spirit travelled through blue air to a cave. Who showed him? He heard the voice of his grandmother. The cave was a burial site. He wanted to get away from this taboo place, but she gently beckoned to him and pointed her long, dark finger. X-ray vision or something. So weird being dead with a pale, pearly skeleton woman, laid among a shell midden. A baby skeleton beneath her crossed bone arms. He saw his heart break. On the woman's skull was a gypsum cap of white clay in a stringybark net.

He knew this grave was two thousand years old. As old as Christ. He was alone, hovering in the plain of forgetfulness. Loved by a young woman, Arana, and she was gone. He dived through clouds searching for her face. It was lonely up there, the cacophony of other blackfellas, out and about, looking for the dawn.

June wondered who else could see Steve levitating above them. The hawk man. The eaglehawk swooping and diving on hot air currents. From way up on mountain ridges. June knew he rose from the earth of warm blood. He scratched at the windows. Wanting always to come in. He had broken wings and was crucified on fences next to fox bodies with red-orange tails. June hid in fear as he asked, 'Why am I here? Let me rest in your arms. Let me make love to Arana with my skeletal frame.' She saw the goonge and his fingers were wrenched off.

The spirits whispered to June as they pounded their feet, bent their knees in bungal djama and danced corroboree. They were covered in ochre and making flames in spirit trees and hid the mourning women, weeping and beating their heads until crimson blood came. Hitting heads with stones. June led her family as they prayed for deliverance. And made another cup of tea in the kitchen, and ate Tim Tams.

***

That morning, June woke up and sunshine flooded the bed. It felt late, and she had no chance of getting up from the lumpy mattress. But she decided she must make something happen. She sensed the local Murries would want retribution, with some internet rallying going on and the young ones checking their Instagram and TikTok pages. A big online response might build, and Arana made an impassioned plea on YouTube and Facebook for justice. Even city media people called the Wilga Land Council for information.

'The sergeant will regret that night,' June said to Merle. 'Behave like that around this town and he might get clubbed on the head one day. Or the fellas might take him out the back and tell him they are rooting his wife. After all, this town is a real Home and Away of intrigue.'

June wouldn't be mute, wouldn't accept that the young men might be flung like garbage into a paddy wagon. The seats speckled with vomit and blood – shit, even. She wanted to speak out, not let others do the talking. A page on the internet spoke about Steve's activism; she hoped to find a seam of bravery and let them see the surge of power inside her. It made her quake.

CHAPTER 33

24th August

An unspoken curfew for whitefellas descended in Wilga, with outside people beginning to arrive. It had once been a town where, years ago, young women could be picked up by cops and raped in the paddy wagon, then chucked out on the gravel road newly swept by Murri council workers with immense black brooms. The event might have set the town alight.

In the Wilga police station, June was back. Flies ran up the dirty windowsills. The grey walls needed painting; they had geometric fingerprints from a crime scene. Maybe some young fella had once tried to escape while a fat copper sat on him?

June asked for a video recording, but the police said the camera into the cell was broken.

If only a witness would appear.

Merle arrived and stood perfectly still. They all suffered from fear or shock. There was usually only fear when it came to gunjies, but this had become rage. As if they didn't care what happened to them anymore. They were not the ones with a dead son, but the death begged for justice.

Steve's family might cry out for revenge, and, being traditional people, they might even blame the Wilga people. The retribution could fall on William, and they might get sung, or a Koradji or galka sent to take the necessary action on their son's behalf. June might have hair stolen from their house and be cursed, or have their clothing poisoned so they went mad. Anything was possible. June was almost peeing herself with anxiety,

'You still haven't given us a copy of your report,' said Merle.

'Give us a break, all of you. Can't you see we are doing our job?' Blackett nodded at June. Perspiration ran across his nose. She saw herself karate chopping his arm off and beating him over his head with the flayed and bleeding end.

William shoved him, and she worried there might be a second murder.

She took William in her arms. 'Okay, thuya-thi, brother, we don't want you arrested too, calm down. Let's do this right!'

He dropped the fist and sat down heavily.

June cried into her hanky. The big office clock ticked, and mozzies hummed in the hot fetid air. The family stared at the dirty lino floor. They felt they were rotting like a dead body laid out on a slab out back of the lock-up. June imagined him again, stiff under the grey sheet, his ghost hovering.

'So,' she said, 'you won't accept our letter asking questions about the death?'

'I've told you,' said Blackett, 'if we want you to make a statement, we know where you live. Sorry you came down for nothing. Procedure.'

'This is why kids break windows, and smash up shops, playing truant and committing suicide. It's permanent victim status,' she sighed.

'Yeah. I know, sorry.'

'Some of our people made a noise when we had the Royal Commission into Aboriginal Deaths in Custody,' she said. 'It looked at ninety-nine deaths throughout Australia, but no police was found to have deliberately murdered an Aboriginal person. Just police lack of care, deception and improper practice. No charges ever laid: is that how it's going to be?'

***

After long bouts of crying, Arana fell to the floor, not able to accept Steve's death. Merle held her flailing arms and tried to hug her, but Arana pushed Merle away and ran out of the house. They all took off after her. Then she was back, weeping. Shivering legs and arms as she heaved from her chest to the top of her head. They swaddled a sheet around her like a baby blanket. Words floated in the air: subliminal messages. Electric currents zooming around the room. Bursts of light in their brains. Making certain that Arana calmed down, they laid her on the couch. Tears and blubbering. They didn't understand how to make it better. Shadows slowly filled the rooms. Arana said she might not sleep at all, ever again. Someone banged on the door but no one answered. A few hours passed. Food grew cold on plates. Eyes swelled with tears. A vast sea of misery. The phone rang every few minutes. Then dead quiet while Arana broke into sobs. They were bone-tired and filled the house with the stink of sad sweat.

'Why him? Why not an old fella like me?' said Jack as the older ones all looked at the floor, thinking similar thoughts.

People arrived at the door: 'If there is anything you need? We're sorry.'

June squatted on the floor next to the fridge. 'I might just have a big glass of vodka,' she sighed.

In the dawn light, Arana sat on the lounge. She saw and felt the goonge while a didgeridoo was throbbing in the dark from her phone. She pulled a blanket over her shoulders and closed her eyes. It was a nameless fear.

A pale figure flew into the room. It had come from the old Country. It possessed the family. It spoke and sang and made them afraid. This spirit with the head and huge wings of a brown and grey speckled hawk with piercing laser-beam eyes. He stared at Arana from high up near the roof; she looked up and spotted him. She sat up in fright as he swooped towards her, then she covered her head in terror and shook and sweat poured down her face under the stifling cover.

* * *

A few hours later, as it got dark, Wilga came alive with Aboriginal people tearing apart the main street. Shopkeepers ran around putting up mesh screens and taking goods to the backs of their shops.

By ten o'clock the place was electric with anger, and June felt nervous that someone might beat up a person unknown or kill him, and that it might be a gunjie. She hoped that the coppers would either get extra help or get run out of town.

Some young fellas had set a car alight, and it burnt with bright-red intensity while they performed a corroboree around it.

The night became more threatening, as the sound of an angry crowd erupted: yelling, smashing glass, rocks thrown. People were dark shadows.

'Black lives matter! No justice, no peace!' yelled the crowd.

There were three locations: the police station, the pub and the street. William, Jack and June moved quickly across the road, ducking rocks. They experienced an exhilaration of owning some kind of power as a group, even if it was just for one night. The whitefellas were cowering in their houses, too scared to come out. Except, of course, for the Gobblers, who rode around the town with bike engines throbbing.

For a moment, June just wanted to climb into bed and cuddle up to her patchwork cover. Just give up and let the young ones fight for Black Lives Matter and their rights. She might take up bowls or macramé. Why not give away her investigating and let Big Cotton have all the water, and just allow herself an occasional loss, enjoy being a defeatist and live on a pension? Let the river die. A duck feather pillow under her head, a sleeping pill.

'Sis, don't worry,' said William. 'Us men will deal with it. You might get a police record.'

'Why don't you have another beer, give you courage,' said June.

William looked stern. 'Am I the man of the house or what?'

'Do what you like,' said June.

'There are other places for us to live,' said William. 'All right. Are we giving up? Want those kids fighting alone? Up you get.'

June put on her Rossi boots and walked out the door with William, her fists clenched. They ran towards the police station and the bikie gang waved at her as she crossed the road.

'Come out and fight, you gubba bastards!' shouted Jack at the cops huddled inside. He hurled a rock at the locked station gate.

***

On the main street, four local police officers lined up for a battle; June was not surprised to see Bundi beside them with a megaphone, calling out for the rioters to go home.

'The coppers have got the numbers, go home!' Bundi shouted.

'Murderers, see how you like it!' yelled William. 'You lot killed Thommo and Steve!' He chucked a rock and hid behind June.

The cops were outnumbered. They retreated to the station gate.

June ran across the road and heard Blackett on his mobile at the police station: 'They're attacking the station and hotel. The customers are on the top balcony, one has a rifle. The publican called. They're smashing windows. I've phoned the next town. It'll be a good hour.'

He then called through a megaphone: 'Go home! I repeat, go home! You are breaking the law!'

June waved at the young Murri men, who were wild and kept up a barrage of stones against the police station gate.

'Don't panic!' shouted Blackett. 'Hold your ground, keep the blasted door shut!'

Johnson held a work phone. He yelled at Blackett, 'Twenty on the way, Sarge. Bloody hell! They're chucking rocks. Get down!'

The landline in the police station rang constantly. Blackett tossed the phone onto the ground.

The publican Arthur shouted across the street from the pub: 'Listen, Blackett, you're the sergeant. Get these Black bastards out of here, they've wrecked my front bar. Someone'll pay for this.'

Arthur ducked the projectiles and ran upstairs to the hotel balcony.

'We're here to protect private property. Law and order,' shouted Blackett.

'Oh, Jesus! Fay Pasqualini's boutique's just been kicked in,' shouted Johnson.

June spotted Bam Bam carrying Fay over his shoulder out of the carnage. Playing the hero. Out of the bushes came Aunty Mary with her boondi stick. June screamed with laughter as she watched Mary bash Bam Bam across his back. He dropped Fay over by the emu cage and flinched away from the old lady. He then ran back and grabbed a rock and stood behind the Black rioters. He threw the rock at the pub and yelled some threats at police.

Arthur yelled through a megaphone, 'This is private property. Get off!'

June heard the sound of tinkling glass, and she watched young ones film the action on their phones. She guessed they were uploading it all to the internet.

'You bastards, you'll pay for this,' said Arthur. 'These renovations cost thirty thousand. Where's the coppers when you need them?'

Jack was below the verandah, throwing big rocks. He grabbed a rubbish bin and jumped on it with some young ones.

'Don't break that, it's not gunna help us!' yelled William to the young fellas.

Riot sounds increased, and Aboriginal people attacked cars and shop windows. A siren screamed and the Murries rushed to the police station.

'You gunjie cowards!' shouted Jack.

'Bastards!' yelled William. 'Run, for Christ's sake, run! Watch out!'

A police officer opened a fire hose on the young people. Among them were Cowboy and Bam Bam holding baseball bats. The torrent of water ripped clothes off backs. It soaked them through.

***

It was midnight when some police reinforcements arrived. They pulled up in wagons with riot gear, looking dangerous. The senior sergeant from another town stood in front, with his torch shining on Bam Bam's face. Blackett faced the bikies, who glared with their jackets soaked.

'I hear you got a riot going on,' said the senior sergeant. 'A few nice rounds in the air will scare the shit out of these cunts.'

'We weren't making trouble, just having a look,' said Bam Bam. 'Better than Netflix.'

'Move on.'

The bikies cursed and spat, but moved over to their bikes and sat on them to watch the riot.

'Better be careful with these firearms,' said Blackett as sweat poured down his nose. 'Can we shoot them? They smashed the station windows, breached the station. Get your arse over there now!'

Videos had appeared on YouTube of the riot. June watched the young ones, who made no attempt to hide their faces, and shot selfies while setting fires. They filmed the police in every encounter.

'Cover your faces!' said June.

'We got you gunjie on filum now!' shouted Jack.

Time stopped, and June felt as if she was looking at the whole mess from the sky. She drifted from intense fear to elation. Her people were fighting back.

Then she spotted them under a yellow streetlight: the black t-shirt boys, holding chains. They stood shoulder to shoulder outside the hotel, alongside Bam Bam. The far-right had come out of hiding. Flying the Nazi flag. They had flaming torches and shook them at the Murries. Sparks flew through the night sky.

June cheered as the newly arrived police moved in on the boys in black shirts.

'Who owns these streets?' shouted the redhead. 'They're our streets. I'm white and that's evil, is it?'

The outsiders with fascist haircuts yelled at the Murri youths. However, the police didn't see it that way, and bundled the lot of them, without their baseball bats, into the paddy wagons with the Aboriginal boys.

The black t-shirt fellas yelled, 'You can't arrest us, we're white!'

Blackett came out of the pub and grabbed the young men by their collars and shouted back, 'You're rioting and breaking the law, and lighting a fire without a permit during a total fire ban. Get into the wagon or I will charge you with resisting arrest, you dumb fuckers. You're not driving into our town to beat up our Blacks, that's our job.'

The police wagon rocked with the added number of men. Inside it sounded as if a fight was going on. The police officers banged on the side with a baton.

* * *

June walked around the riot and spotted people running down the road carrying fence pickets. She wanted them to stop the violence. Among them was Duckie, carrying a machete.

'Hey, Duckie, come over here!' June shouted, but the woman ignored her and kept moving with the boys. She looked furious.

The last sight of Duckie was her figure against a burning car. She was smashing a headlight.

'Watch out, there's the mob from the mission,' said Jack. 'Wild fellas. Gunna chew up the gunjies and spit 'em out.'

William arrived in the middle of town, where some police were standing on the corner. He saw their riot gear, but sauntered over, nice and calm. He watched Arthur outside his pub.

'Hey, the young blokes are stealing the pub's liquor. Free giveaways, is it?' yelled Arthur.

'Piss off, Arthur!' yelled June.

By two o'clock in the morning the pub was still under siege.

'Go to the park!' June shouted while the publican threw stools from the hotel balcony onto the rioters. 'Please, everyone calm down and go to the park. A man at the hotel's got a rifle. Please move away!'

All around June the noise of shouting intensified, and she huddled in the relative safety of the bus shelter next to the hotel. She saw young men being hit by batons and kicked by police. They were beaten down and some were badly hurt.

June felt a power. She snapped and ran down the street towards the riot. A Black hand picked up a stone and the glass from the bar smashed over them in a shower of blood. Her voice was hoarse with begging the mob to go home before they got arrested or killed by a rifle from some ute. Farmers had 202s sticking out windows. She ran with the mob because she wasn't going to take it anymore. She would punch in Blackett's head; she would find the records that proved what happened to Thommo and Steve. It was her goal in life.

Arthur spotted her and yelled, 'June, come over here, you can hide in my bar.'

'I'd rather be with my people!' June shouted back, and saw him pull a table to block the door.

'What, I'm a white supremacist now, am I?' yelled Arthur.

'You're white, Arthur!' yelled June.

'That's racial prejudice. Some of my best friends are Greeks.'

There was Duckie running towards June with a look of excitement and terror. She bounded across the road and took June's arm. 'Come and help me start another fire at the cop shop!' said Duckie.

'Don't be mad! You'll get in terrible trouble!'

'Don't give a shit anymore, but! I'm getting revenge on those pricks. They took my husband, but, and I don't care if I die now!'

'You will lose your son once you get thrown in gaol. DOCS will take him and foster him out. Don't be stupid – come here and stand behind me. And for God's sake throw away the weapon.'

Duckie collapsed on the concrete next to June and leant against the wall. She cried and June stroked her hair.

* * *

By early light, the rioters were either locked up or in paddy wagons on their way to other towns. Blackett pushed Johnson aside and unbolted the door and they peeked out at the office covered in upturned chairs. 'Tidy up, will you!' said Blackett. 'Need to look like we have it under control. The senior sergeant will want to talk to us.'

'We were scared of them, and I reckon they deserve to be angry; I would be. I don't understand how a few punches could do that,' said Johnson.

'Now you know, I only laid in a few … Careful what you say,' said Blackett.

* * *

It was morning, and Johnson and Blackett were back in the police station. They waited to explain themselves. They kicked open their snack dispenser and ate all the Mars bars and chocolate milk.

'You're wrong, Sarge. I should never listen to your crap. I know what's right,' said Johnson. 'That man had his whole life before him, and I stole it.' Johnson licked the chocolate wrapper. 'You were in on it too. His head made that sound, a crack. I murdered him, and nothing you can say will change that. I must face myself. I won't crawl away like a dog. I am guilty. Or did you go back and hit him again? I didn't see it. I will leave the force. Well, I'll probably go to prison for murder. I've ruined it all. I will never forgive myself. I'm going to confess. Don't stop me.'

'Wouldn't dream of it, son. Your call. I don't give a shit about your morality!'

'All he wanted was a phone call, for Christ's sake. One call.'

'Don't blubber. I will bloody punch your lights out if you blubber.'

'I can't help it.'

Johnson thought that the sound of sirens coming up the road were for him and not the rioters, and they would take him away and throw him in a cell. Steve's face was whispering in his ear. That dark eye in his eye, not ever blinking. Steve was looking at Johnson, really seeing him for what he was. A poor stupid whitefella with no self-control. No one had ever looked at Johnson like that.

A wedge of deep black sky shone through the bars. The sound of the sergeant beside him breathing his deep rattling breath full of mucus. The incident an appalling mistake.

Blackett shook him by the shoulder and said, 'Pull yourself together. They're coming.'

Johnson knew that they were coming to take him away. He heard Steve calling, and that eye was still inches from his own watery eye.

The sirens were close now, and the lights washed and flashed through the station windows like the fireworks on the Sydney Harbour Bridge.

'It's coming back to me, how it happened,' he said.

Blackett said, 'Oh, yeah? Take off your belt, son. They will want your weapon.' He held up his hand, and Johnson looked at it as if it was on the end of a plastic trick arm. The belt dropped in the sergeant's hand. Johnson was lighter, as if this load had floated away.

'Gotta make a full and totally truthful statement,' said Johnson.

'Whatever you want, mate,' said Blackett. He sounded old, needing sleep, as if it was all too much. He wanted the day to end so he could get into bed. He wanted to retire from active service as soon as he could get his pension, maybe buy a big bike like Bam Bam's.

* * *

Merle was in the kitchen at eight o'clock in the morning. She made tea and sat in silence. She even cooked pancakes with lemon and sugar, all while the world was going mad.

Meanwhile, June saw Jack hobbling along the street near the house. He had taken something from the yard. He would destroy the police station and kill Blackett. She crossed to him. The street was quiet now; people had gone home. Jack jumped up and down with a hammer behind his back.

'Jack, don't be a mad idiot. You go home,' she yelled. She waved at him, then realised she would have to crash-tackle him. She tried to recall the actions of tackling below the knees, but she might hurt him or do her own knee in. Still, she had to save him from himself. Her chest pounded, and then she had him by the legs. He raised the hammer like a serial killer. Down he fell.

'What the hell?' he said.

'I should let the gunjie flatten you with a cattle prod,' said June. 'Give it here.'

She escorted him home.

Merle sat by Uncle Jack and handed him some tea and a pancake with lemon and sugar. June snuck out the door and went back downtown.

'I wanted golden syrup,' said Jack. He licked the spoon. 'Our Arana, she got something special. I'm real proud of that girl. We can fight for that young fella. Never had me own kids. I'm finished, all worn out. The gunjie shits have kicked me to hell. Now, I can give it back for her. We don't have to say yes boss, no more,' he said.

William came back from the rioting street. Dust and sweat covered his face and he looked elated.

'We weren't gunna stick around when they're smashing windows,' said Jack. 'Those gunjies were pissing their pants.'

'Yeah, they wouldn't listen,' said William. 'Young fellas tore up the pub. Don't know why – they'll have nowhere to drink tomorrow. Arthur kept saying it was closing time! June gave him what for.'

'Is she all right?' said Merle.

'She was riled up and wants to fight them all.'

'The system's stuffed, from the Rum Corps to now. It's the same bad story. They have locked us up or shot us, and now they've stolen the water,' said William.

'Or electroshocked me,' said Jack.

'Don't, for God's sake, let him get worked up!' said Merle. 'We got enough trouble here with our girl,' said Merle.

'We know, old fella, it was bad,' said William. 'But now we got to keep our heads.' He looked out the window.

Merle put up her finger to hush them. 'Quiet, you two, she's asleep. I gave her a sleeping tablet.'

Jack opened the window blinds. 'Eh, Lord, they burnt a car. I saw the gunjible run! Go get him, knock the goona out of him! Bet they bloody get away with it,' said Jack.

'There should be one law for all,' Merle said.

'Yeah, Blackfella's law,' said Jack as he paced back and forth.

Merle pulled him back to his seat.

'You look around Australia, all Aboriginal ghosts,' William said. 'Wish they'd bring back tribal law. I'd spear the gunjie and put him on the ant's nest.' He buried his head in his hands.

Merle put her arm around William to comfort him.

* * *

In the warm morning sun, pink light shone on June's face. She sat on the road near a paddy wagon as several police officers arrested people they had seen destroying property the night before. She sat among the broken glass and two big gunjies picked her up.

'The police killing our boys in custody. We want justice!' said June. She didn't struggle, but passively resisted, just like Gandhi. They shoved her into the bull wagon. She saw Bundi standing with Arthur, watching. The Mission fellas banged on the metal side, and she slumped down on the seat. The noise gave her a headache. Merrigan dogs barked, and she looked around the wagon to see who else had been arrested. A young boy from the Reserve, crouched big-eyed in the corner. June thought she would never forget what whites had done to her people. It was written on his face.

Smashing and looting. It might make people realise what the economics of the town were made of, with no businesses owned by Aboriginal people getting smashed up. The adrenalin rush, wanting to break a world that didn't allow them a place, except at the bottom of society.

CHAPTER 34

10th September

Arana was silent at first, but as time passed she stood by the window looking out and thought about what had happened and had a fear the memory would never go. Springtime and her lover gone. She'd known Steve would be snatched away from her, that he was too good to be true. She was drifting away along the riverbank in a mist of sadness, immune to kind talk.

She stood in front of the 1970s bedroom mirror. She looked in the large disc with deckled edges over the dark wooden dressing table. She fiddled with the cut-glass powder bowls and perfume sprayer. No powder now, just bobby pins and an old can of Gossamer hairspray.

It was just like the night of the ball, with the spraying of her elegant hairstyle. She took up the bottle of skin moisturiser, Oil of Olay. It smelt of sweetness and she dabbed some on her cheek.

Her face shimmered in the mirror. She squeezed her eyes shut and willed herself to see Steve alive, laughing next to her. The nape of his neck with the soft black curls.

She imagined him graduating from law, waving to them from the podium. He would have worn the red, black and gold sash. He might have been her constant companion, her thread of life. Her guide to his true homeland. Now, how could she meet his family? Did they blame her for his death? Would they want to punish her? A payback of some kind? Arana shook her head. If only she had not spoken out in the street. If she had been submissive and scared instead of defiant. She thought about that night and went over the details. She had been reckless. If only Johnson hadn't wanted to show off in front of her . . .

* * *

The warm air smelt of wattle and boronia flowers. Young people were frisky. The family gathered around the kitchen table, drinking tea and eating Arnott's biscuits. Fighting over the last Monte Carlo. June said Arana didn't want to hear anyone say that Steve might have committed suicide; she went crazy if someone hinted at it. June agreed: as far as she was concerned the police were murderers, and involved in killing Thommo, too.

Merle tried to reason with Arana because everyone was scared that she might hurt herself. They had cousins to take turns to keep an eye on her, and Arana couldn't even go to the jillawa without someone going too. She yelled at them to go away, opening the bathroom door and screaming. For June, the deepest fear was that this young girl could be developing problems with mental health, like Uncle Jack. If Arana fell into depression it would be a tragedy.

June had a great fear of being alone in her old age. The spectre of isolation haunted her.

'We need some babies to keep us going,' said June.

'You'll always have us,' sighed Merle.

***

After a while, June called the local doctor. Ian was a nice young fella who grew up in Wilga, and he had a wife who was Indian and sweet to all the Murries. He came around to talk to Arana, as she hadn't eaten for days, and she stayed in a bedroom sobbing to ghosts. He sat on the lounge and held her hand while he kept repeating to her and the family that she would recover, that she was still in shock. The doctor offered medication, but Merle shook her head.

'She will come out of it. She needs our love, not drugs,' she said.

The doctor said to watch Arana, and counsel her as often as they could, especially William and Merle. Merle reckoned her lips would drop off from endless talk while Arana just rocked in her pyjamas, glued to the TV. After a while, June watched TV all day with her and made lots of tea and got quite caught up with the Kardashians. They sat on the lounge in dressing-gowns and nibbled chips or combed each other's hair.

***

Arana stood in front of an open window in the evening with rain pouring on her face. Lit up by flashes, her hair flew: a banshee. June slammed the window shut and looked at Arana. She was unmoved, her face pale, as if a vein had opened, life draining away into the tumultuous night sky. The lightning flashed again, and June spotted a figure in a dark coat by the gate. He waved; it was Bundi. She pulled the curtain shut.

The sound of knocking and she heard Bundi calling out. William opened the door. June shook her head, but he was in the room wiping down his rain jacket.

'Just came round to say goodbye to old mate here,' he said. 'I got a job in Brissy, big pay, for an international cotton growing consortium. Don't frown, Will, they need to understand our cultural perspective.'

'Yeah, big pay, eh?'

'Not why I took the job.'

'Whatever brings in the walang, eh? Got yourself a new car, I see.'

'Don't be like that, brother. Hey, June, I want to discuss you coming with me. Start up an agency up north? Come for a spin?'

June twisted her mouth; she wasn't sure about anything anymore. She nodded, and to Bundi's amazement she put on her jacket and ran to the bedroom to spray perfume on her neck. She walked out to the new hybrid vehicle with him. They drove off to the outskirts of town and parked by the levee bank in a quiet spot by coolabah trees. It was romantic.

He grabbed a bag of cans and offered her a beer.

She shook her head. 'You know I hate drinking!'

He sighed and threw the bag of beers over the back of the seat. They sat in silence for a few minutes. He laid his head on her shoulder and hummed.

'Could you love me a bit, though? I promise to be gentle?' He quivered beside her and traced her lips with his thumb. His hand slid to her breast. She sighed.

'It's over,' she said. 'It won't work with us. You're going in a direction of politics that I can't stand.'

He ran his hand up her black skirt. She shivered and turned her head away and pushed his hand back into his lap.

'You're a hot woman!' he said. 'But I respect your choice to say no. I've done the training about women's rights.' He grinned but kept his hands to himself.

A knock on the window. Sergeant Blackett shone a fierce torch in her eyes.

June slid down the seat and hid her face in her arms.

'Hey, are you doing some investigation work?' Blackett said. 'You could be arrested for obscenity!'

'Sorry, Sarge,' said Bundi.

'It's public indecency, unlawful behaviour.'

'We'll be on our way, thank you for the warning.' Bundi started the quiet engine.

'Give her one for me,' said Blackett.

'Drive me home,' June said. They drove off.

'Anything you ever want or need, I will be there.'

'Sure you will.'

He drove June back to her house and escorted her up the steps, where he kissed her sweetly goodnight.

Bundi winked and left her at the door and June tiptoed inside. William smoked in the lounge room and twitched the curtains to look out. He growled as he watched the car drive off. The new white Lexus zoomed away in the night.

'His old car was a Pajero,' said William. 'Means "wanker" in Spanish.'

CHAPTER 35

13th September

Some weeks after Steve's death. Out back in the Ngyiampaa and Yuwalaraay country, a Toyota police car cruised on a hot, quiet-as-death afternoon. From the grey sand dunes they could smell the dying Darling River.

Sergeant Blackett and Constable Johnson were 50 kilometres from Wilga, on the way back from a farm where a break-in had been reported, but no one was home when they turned up.

The car's bull bar was covered in dripping red dust. They drove over rough unsealed roads in a wagon that had held desperate men and crying women on blood-spattered cushions. They passed silently through a vast cattle station, the biggest in the world. A monument to what whitefellas could steal and keep, and wreck with abandon. *This was God's Own Country*. Johnson kept these discomforting, subversive thoughts to himself.

They drove through gidgee, weeping myall, desert kurrajong, wild orange, quandong, saltbush, leopardwood and spindly ghost gums. They swerved across newly cleared land of upturned trees and piles of timber.

'Farmers have cleared the bush,' sighed Johnson. 'No wallabies now, lot of feral goats. Even the Bourke's parrot is endangered. Pretty orange bird.'

'Just rubbish trees. But the rain's bringing green shoots.'

Spindly wattles leant towards the vehicle. Rusty sheds from abandoned farmland littered the stones, then the sun set in a blazing pink. It was quiet; nothing much to hear but the occasional fart.

'Sarge, that smells bad.'

'Suck it up, son.'

'I had a dream about that Aunty June. She came after me on Bam Bam's bike with a huge dead Murray cod fish, chased me along the road.'

'She's a formidable Murri investigator. Dangerous, better stay out of her way,' said Blackett. He had been thinking nonstop about her. She had looked him in the eye and her eyes drilled into his head. Everything was different. It was written in the contempt she spat in his direction. She knew! He felt sick.

'Her niece is so beautiful,' sighed Johnson.

'June told me that old Black woman Aunty Mary has pointed a bone at us, because we weren't suspended from duty,' said Blackett.

'What bone?'

'Not one from Red Rooster,' laughed Blackett. 'Not finger lickin' good.'

'You scared?'

'Nah, garbage.'

'Sure,' mumbled Johnson. He gazed across the dark tree line to the high riverbank. 'There's what's left of the Darling River. I heard it's rising.'

'One moment a brown dribble, then after a few weeks of torrential rain up north it turns into a great sucking monster, all right. Tear up everything in its path,' said Blackett.

'Not here, yet.'

'Crashing lightning, set your hair on fire,' said Blackett. 'That June is taking us to court. Bloody stressful. She's got an Abo barrister to represent his family.'

'That's an Indigenous person.'

'Shut the fuck up and listen. They've got big guns, want us convicted. Murder. It won't stick. The union will fight. Twenty years in prison,' said Blackett.

'Yeah, our union, no worries,' said Johnson.

'I'd be eighty.'

'I'd get sexually assaulted.'

'Squeal piggy,' snorted Blackett.

'Don't.'

'They would say you had a nice bum. "Buns", they call it.'

'Time off for good behaviour,' said Johnson.

'I'd neck myself first. She's a Black vengeful bitch.'

'She's always been nice to me.'

'Nice is a cold beer. She's dangerous, your worst nightmare.'

Hours passed as the moon rose, and the men drove on beneath its luminous glow.

Up north, the serpent dived into the grey swirling river, and its mouth gaped, and it tore along the rivulets and quiet creeks. The rivers gushed and squealed into life.

'Ghostly galleon, tossed upon foamy seas ...' said Blackett.

'What'd you say, Sarge?' said Johnson.

'Ah, forget it.'

'Pity about the unsealed road – we were hitting a hundred and forty back on the bitumen,' said Johnson.

The vehicle lurched over rocks. It was getting dark.

'You idiot,' said Blackett. 'Drive slow on this road. We don't want a blow-out. If we do, you'll be changing it.'

In the yellow headlight glow they glimpsed old tyres piled against barbed wire, thorn bushes, wooden posts. Strung along the wire was the crucified corpse of a wedge-tailed eagle, its great yellow beak bent to kiss the earth. A skull and crossbones flag was flapping next to an abandoned caravan.

'It's like Bam Bam's place,' Blackett said.

'He's a murderer, I reckon.'

'He's all right.'

On a disused highway, they rattled along past an abandoned shack with a rusty sign, 'BP Service Station', in peeling red paint. Johnson gazed at the forlorn building, and he wondered whose life was spent out in the middle of the semi-desert of broken trees and dreams. He noticed a shadow move inside the ruin. He shifted in his hot, sticky car seat. It stuck his shirt to his back with sweat, but he experienced uncomfortable edginess in a way that had little to do with heat. He chewed his nails and felt a deep unease, and a rising, a foreboding, a sickness that came in waves and made him feel dried-up and lost. It had been that way for weeks.

'Wouldn't want to fill up there,' he said.

'Hope you filled her before we left.'

'Plenty.'

'I need a piss. Pull over.'

Blackett had not been sleeping well ever since it happened. Waking some mornings to find handfuls of grey hair on his pillow. Scalped in the night. He couldn't even talk to his wife. She didn't even let him sleep in their bed. The bitch. He longed to be held with compassion. Or even the occasional conjugal root. He would even give June one, if he had the chance. But he wasn't big on pity. It was best locked up cold like a vault of steel. The tightness around his chest a vice. Maybe he would have a heart attack any day now; that would teach them. Shut that steel trap of teeth and wait until the noise died down. They would stop asking questions after a while. The police union would stand beside him, would look after a man with his impeccable record.

He urinated in the dust and zipped up his pants. They drove off.

* * *

Some hours later, Johnson drove on in the dark and leant forward to see through a grey mist lifting from the hot road. The desert strewed stones across the landscape like giant toys, and dying red gum trees hovered all the way to the horizon. The men twitched and suffered sensations of total inexplicable anguish that grew and grew until they rounded a bend in the road. And then, there it was.

A ghost in white ochre and blood burst out in the headlights. Its semi-naked dark body trembled, and its head had a gash that flowed with gore. It had eyes that shone like powerful light beams. The semi-transparent image shivered and pleaded with hands outstretched, and its mouth seemed to be crying for help. Its fingers trembled, a drop of blood fell to the earth, and its eyes wept and stained the sand.

The police officers gaped with wide-eyed gunjie terror, and the brakes slammed with a screech. As Johnson gripped the wheel, he yelped and swerved off the road, and the vehicle crunched into thorn bushes and rattling gravel. It stalled, and Johnson looked at Blackett, eyes staring, his face transformed to a white mask. He tried to restart the car; the starter whined, and he turned the key, again and again, and it crunched and twisted and died.

'Start, for God's sake!' he cried.

'You'll flood the engine,' growled Blackett.

'Start, you bastard!' Johnson cried even louder. Sweat dropped down his face. He gulped, the car jumped, they drove off. They thundered onto the road again with a squeal of tyres.

'She probably sabotaged the engine,' Blackett said. 'She said we didn't investigate the greenie murder.'

'Don't think so, Sarge.'

They drove, their heaving breaths filling the vehicle with a fearful misery.

'What the hell was that?' said Blackett as he turned around and looked out the rear window. He ran his fingers through his hair and laughed. Perspiration trickled down his thick neck.

'Christ!' said Johnson. He giggled nervously, and licked his lips. His teeth stuck dryly in his mouth. He swallowed and nearly sobbed. His heart raced, and his hair stuck upright like lightning had struck him. 'My god, what the hell was that?'

'Just bloody drive. Not a bloody word,' whispered Blackett.

Ten minutes passed in grim silence. Johnson looked in the mirror, like he had stuck his finger in a light socket.

'I don't bloody believe it. Fuck, fuck, fuck!' said Blackett.

'Shit! I saw it, too.'

'What?'

'Ah, nothing, Sarge.' He paused. 'Must be goin' crazy. Jesus!' The sound of his short puffing breath was loud.

'What do you think it was?' Blackett's voice trembled.

'Must've been dreaming. Nah.' Johnson paused. 'Just ... back on the bend, I reckon ...'

'I know,' whimpered Blackett. 'That witch sent a letter threatening to expose me about Steve Allunga.'

'You tore it up, eh? Did you see something on that bend?' Johnson's voice was as small as a mouse.

'Get out of it!' Blackett boomed. He scrambled for a cigarette, and his fingers shook as he cupped two hands to light it. He flicked the match onto the floor among the Mars Bar wrappers. His heaving chest couldn't catch a breath. He coughed and pumped his ribs with a fist. Tumbling pictures of a man dying.

'You saw it, didn't you?' Johnson said. 'We didn't imagine it. Bloody strange, eh?' His nervous laugh punctuated the quiet.

His boss tried to use a calm, controlled voice, but he shook and bashed at his knees. 'Yeah ... Could've sworn there was someone ... something. I don't believe it.' Blackett paused.

'What? Say it. A blackfella? And he was hurt bad, standing in the road asking for help, blood all over him?'

Johnson turned to stare at Blackett, who appeared to be in shock, his eyes wide and staring.

Blackett went on. 'Hell, all me hairs are on end! A bloody ghost! Never ever in my whole life have I ever seen anything like it. It was a skeleton with a pointed finger, skin. I need a drink, a Valium.'

'It was him, wasn't it?' said Johnson.

'Don't be stupid,' said Blackett.

'I was gunna stop the car and get out.'

'No way I was getting out.'

'But he was there.'

'We come around that bend and I *knew* he was going to be there. Bloody hell, gives me the creeps.' Blackett laughed, filling the car with ugly noise.

'It sounds weird,' Johnson said, 'but I started getting this, I don't know ... awful sensation ... about twenty kilometres back.' He paused. 'Far out!' And paused again. 'How far to town, Sarge?'

'About sixty kays. Ghosts. Nah ... Jesus Christ! Hope it won't be following us around. Gawd, I been working too hard. I need a bloody break. I get depressed about this kind of thing, bloody black dog!' Blackett gasped for air; his shirt collar was suddenly choking. A pulsing purple vein was going flat-out, but he couldn't see that.

When the younger copper looked at him, it was if the old man would expire from a heart attack, and then Johnson would be out here on the road in the middle of bloody nowhere, pumping the flabby chest and trying to find the solar plexus and breathing into that big-lipped mouth with the false teeth with that annoying click they made.

'Poor bastard,' Johnson said. 'All that blood! He was begging us to stop! Like it was a silent movie. Gee, the guys at the station aren't going to believe this.' He reached for his Extra spearmint chewing gum on the dashboard. He offered the sergeant a piece and neatly folded the wrapping and placed it carefully in the ashtray. 'I was going to tie a tourniquet around the wound. You know, stop the bleeding. Or use my hand to apply firm pressure.'

'What, around his whole head? Are you mad?'

'No, a pressure bandage or something like that. Not too tight. I must read up on my first aid; I can't remember. You know, we should redo that course every year.'

'He's dead,' murmured Blackett.

'He might have been hitch-hiking. We gotta go back and help him, or call an ambulance or something. He could bleed to death out there.'

'Right, Mr Superman, want to jump into that dead tree and call? No reception for your mobile out here, sweetheart. Don't make an idiot of yourself, son. Keep this quiet, eh? And don't let that gin hear about it, or else!'

'That's not very polite, that word, "gin". Maybe we can go back to Bam Bam's place – he has a sat phone.'

Both the officers stared out the windscreen in dumb incomprehension. Blackett could see a hurtling lightning bolt full of questioning faces.

'I thought maybe we should report it, you know, like extra-terrestrial sightings and other related stuff,' said Johnson.

'Are you for real, Johnson? Were you born yesterday? Can you just shut up and drive? Can you do that for me? Extra-terrestrial, my arse.'

'Well, explain it, then. What was it? Who was it? Who is responsible?' murmured Johnson.

'No one is responsible. Say it to your effing self. Do I have to spell it out? It's bullshit. Don't fall for her propaganda.'

Quiet tension hovered in the heat.

Johnson whistled. 'He was there ...' he whispered.

Blackett shifted in his seat and turned to face his constable. He poked him on the shoulder. 'Now, let's look at that dumb statement, shall we? You signed the legal document about it, is that right? Is that right? Answer me!'

'I signed it,' said Johnson.

'Your bloody name on the New South Wales Police formal report: you saw nothing. We saw nothing.'

'When? Then? Or just now? Because I swear I saw it. You saw it. Never ever seen anything so scary . . .'

They shivered in unison and a heavy silence descended.

'Just drive, and keep your attention on this road. Kangaroos,' said Blackett.

'Should report all incidents,' said Johnson.

A rolling ball of white light swerved across the road and under the car. They didn't speak, but each noted the strange phenomenon. Min mins.

'I bloody well should put you out of the car and run over you myself. What a loon.' Blackett's face distorted in an angry grimace. 'Who is responsible? Well, we know the answer to that, don't we, sweetheart?'

The vehicle rattled against stones and plunged into thick white bull dust and stalled again. It soon bogged them, the wheels turning in the deep ruts.

They didn't want to get out in the darkness with its creeping spirits. Blackett finally climbed out of the car and pushed while Johnson revved the engine.

The sergeant slammed his fist on the roof and shouted, 'Go, go, go, damn you!'

Johnson looked in the rear-vision mirror and saw a large figure with a torch light standing behind Blackett in a glowing ghastly light. It was Bam Bam, and he held a rifle and aimed it at their vehicle.

The sound of the river in the distance was a menacing roar. The lightning came, and clouds raged across the teeming sky. The breaking drought and rise of Waaway the great serpent.

'Sarge! Behind you, there's . . .' Johnson called out with his face at the open window.

'What?'

'He's there, it's right there!'

Blackett swerved around with his hair sticking up. There was nothing, just a hollow space and eerie expectation. Darkness and emptiness. 'Shut up!' he said.

'It was the bikie!'

Blackett heaved the vehicle with his shoulder and slammed his hand down on the back of the wagon. Johnson hit the accelerator, and the wagon was out of the deep rut, leaving a cloud of white dust over his boss. Blackett brushed off his clothes and climbed heavily into the vehicle. They drove off in a quiet hush. Johnson's hands were shaking.

'What did you say back there?' said Blackett.

'I said: it's Bam Bam.'

'Didn't sound like it.'

Johnson clenched his teeth and his eyes squinted in the mist. His heart beat like it would take off down a hellish runway.

CHAPTER 36

16th September

Merle, Arana and June were standing outside the house, under the stars, in the dark, watching the peaceful night, when a snowy owl crossed over the house. They ducked down. Those owls brought bad news. The women pressed up close to Arana; she could feel their warm, soft bodies.

'I'm scared that I'm going mad,' said Arana. 'I should have been courageous. I should've followed the wagon to the police station. I guessed who killed Thommo. I could have done something. Not just walk away.'

'You're only a girl,' June said.

'I could have saved him. I wasn't brave.'

'Every sane Aboriginal person is terrified of the gunjies,' said June. 'They lock us up, kill us . . .'

'I wanted to see his body. They wouldn't let me see him,' said Arana. 'I wanted the truth about his death. My heart is breaking. He's gone, but his shadow is like a wind rushing all around me.'

June wanted to soothe away her anguish, but Arana was trembling like a leaf. Part of being a Murri or Koori: having that terrible load of history to carry about like a load of stones on your back. It never bloody went away.

They heard a yell, and Arana stood before them with glowing eyes. Merle took Arana firmly by the shoulders and held her tight. Arana pointed into the bedroom and her hand shook.

'What you're seeing is his spirit,' said Merle. 'Ask him what he wants; he's got a message for you.'

'But I'm terrified!'

It's all right, darling, we will take care of you,' said Merle.

'He can't hurt you,' June said. 'It's okay to see a ghost, we see them all the time. It's like getting on the phone.' She hugged Arana. 'I'm going to ask Fay Pasqualini for some of Blackett's hair so I can give it to old Mary. She can put a curse on him for sure,' said June.

Merle nodded her head, but there was something else to talk about, and she swallowed and stood up and placed her hand on Arana's back.

'What was in the coroner's report?' Arana said.

'The coroner ruled that there wasn't any evidence of violence,' said Merle. 'They took the police statement. No inquest ordered. He reckoned that it was suicide.'

'That's impossible!'

'The police were commended for professionalism,' June seethed. 'I can imagine them lying. That Wayne couldn't look at anyone. The *Wilga Gazette* reported that Steve died in the sergeant's arms. Cow cockies' paper.'

'I really wanted to go to the funeral,' Arana said. 'His people up north gave him a good burial. His coffin all painted up with his Dreaming totems, and they had a church minister, but the family performed the mortuary ceremony. And not a white man's coffin.'

* * *

A few hours later, very late, June heard the front door close, then footsteps outside, but she was so terribly tired, so she turned over and pulled the pillow over her head to sleep. She knew Arana now walked the streets at night.

It was not until morning that June and Merle realised Arana had not returned, and they panicked. June rushed into her clothes, calling out for Arana, while Merle contacted all the relatives.

Arana, meanwhile, was walking in a sudden rainstorm out on the plain on the edge of town. Black clouds billowed overhead and lightning flashed purple. Water ran down her body. She had paced half the night when she saw something standing on the road. She called out over the tossing trees and clouds and peered through the darkness in the direction of the river, then she realised that she was seeing Steve's ghost. He stood in the sky, but the vision faded in and out. Arana wasstunned, and buried her face in her hands.

'Something told me to go along this road … looking for this moment,' she said out loud. A loud clap of thunder shuddered in the air. Arana dropped to the ground, then laughed with relief. 'I thought … I thought … I'd be too scared to do anything, eh, Steve … I'm still shaking … You came in a dream … You knocked on the door, you kept knocking and I got up … I knew it was you … Could feel you through the door … The walls … Melting like honey.'

The storm rose again. Her voice fought over it. 'Steve! We couldn't see across, it was so dark … The water was wild … Jeez … Steve, I tried … I pushed that boat, which was a coffin, over the water … You lay there in the bottom.'

The storm abated, and her emotion was spent. Arana's voice was almost a whisper. 'Steve … I got you there.'

* * *

It was a warm September morning in Wilga. The town shone, with new white walls on the courthouse that the council had repainted. The road was deserted except for the occasional mother with a pram and toddler. The street sweeper was spraying water down the road to keep the dust from blowing into shops.

Outside the Sweet Trolley cake shop, Constable Wayne Johnson stood and looked at the arrangement of cakes. All lovely, with doilies. He needed something really nice. He was in his civvies and waiting for the transfer to come through to Broken Hill district.

The bell over the screen door tinkled.

'Morning, Constable,' said the cake shop owner. 'What can I do for you today? Same as usual? A beef and mushroom pie with barbeque sauce and a vanilla slice?'

'No. I want something nice for a special occasion. That sponge cake with strawberries looks good.'

Once he had the cake, he held the box out in front of him and walked the couple of hundred metres to June's street. He stopped on the corner and breathed in deeply. He was thinking about what happened; he hadn't been able to get it out of his mind. He felt like he was going crazy. And, meanwhile, Sarge was losing his marbles on leave.

***

Arana had seen Wayne parking outside the house for weeks. No one had told him off. Every time she looked outside she saw his police car drive slowly past.

There was a knock on the front door, and Arana saw June open it, and she stood there, not smiling, just standing with her hands on her hips. Arana hid behind a door.

'Hi, Aunty June,' mumbled Johnson. 'I wanted to pay my respects.'

William nodded at Arana, and came to the entrance and stared at Wayne. 'What you want? It's not a good time, bud; you shouldn't be coming here,' said William. 'I'm just about to go to my Gamblers Anonymous meeting; can't talk now, fella.'

June stepped aside to let Wayne into the house, but her face was thunder.

William stood in front of him with his burly arms crossed.

Arana called out, 'Tell him to get!'

Johnson said, 'I brought you a nice sponge cake. Better eat it soon 'cause the cream might melt – look.'

'What do you want?' hissed William.

'Come to say sorry?' Arana yelled from behind the door.

'My mum said I should say something. It's been the hardest thing to get up the guts to face you. The force will transfer me soon, so ...'

Arana walked out in her dressing-gown with hands pressed in her pockets. 'Can't wait!' she hissed as she pushed her uncombed hair from her eyes.

'Spit it out, son,' said June.

'About that night. Sometimes we want our own survival.'

'You did,' said June.

'No. You're accusing me? I'm a liar, am I?' Johnson squirmed.

Arana felt hard and cold. 'We've heard pigs lie all our lives. We're waiting.'

June looked at him, he was trembling with his hand gripping the table.

'She called me a pig,' he said.

'She doesn't mean to hurt you,' said June.

'Hurt him? I'll bloody well hurt him, all right!' shouted William. 'Out, out, before I punch your lights out!'

'Calm down,' said Johnson, his hand up in supplication.

Arana grabbed him by the collar and manhandled him towards the door. She looked suddenly strong.

Johnson gripped the doorjamb. 'I'm sorry, but you must accept it was an accident. They train us to react when we are under attack, and he hit me first. I'm a well-trained police officer. I didn't know that the sergeant would go nuts and bash him.

He hit his head. Then, you know, he killed himself later on, in the cell, like, I didn't know that could happen, he was so quiet, drunk . . .'

Arana's face transformed, and she shuddered. 'He never drank, ever.'

'Ah, well, he might have told you that, but . . .' said Johnson.

June stood at the front door with her hand out, motioning for him to leave. Exhaustion swept over her, and she saw him as a pitiful figure.

'Go to your room, Rara,' said William. 'You don't want to hear this fella.'

'No, I do!'

'It was an accident,' Johnson said. 'I swear to you, I would never hurt him on purpose.'

'He didn't suicide,' said Arana.

'Well, it looked like it, but we had no CCTV; it wasn't working,' said Johnson.

The air prickled with tension, and the sound of hard breathing filled the room. June thought Arana wanted to throw him down the steps.

June, William and Arana stared at Johnson. He stammered, then put up his hands in supplication towards Arana. June opened the front door and indicated it was time to go. Wayne looked anxiously at Arana and tried to smile.

'I got you a nice cream sponge, Rara,' said Johnson. 'Couldn't we make a cup of tea? At least you have to give me credit for facing the family. I'm not a coward. You sure you wouldn't just come for a pizza before I leave? It's spring.'

'Out,' sighed June.

'I thought that seeing as you are single again, Arana,' he said, 'you might want to come to the pictures with me? The film club is good.'

'You punched him out and watched him die!' said Arana.

'No, never, I never – Sarge did it! I didn't mean to hurt him, it was a mess, the Sarge was yelling at me, and I don't know what happened. I can't remember details – you gotta believe me, it was an accident, I swear!' he shouted.

Arana slapped his face again and again. Wayne flinched.

William intervened to take her hand away, then he walked him to the doorstep. 'Better go, bud. Tell your mum we're really sorry they mixed you up in this. Your mum is a fine woman.'

'But what about the cake?' said Johnson.

'Take it,' said June as she bundled the box into his shirt front.

'No, couldn't, it's not good manners.'

'What can I do for you?' whispered Wayne as his hand reached for Arana.

'Shoot me,' said Arana.

'Quiet, lovey,' said June.

'Get out! Out, out!' shouted Arana. 'We never want to see you ever again. Go away and die!' She began to sniffle.

'Forgive me?' said Wayne. 'Just say you do, even if it's a lie.'

Arana was blank. A tear fell down her cheek. 'You took him from me.'

'If I died, it wouldn't bring him back.'

June imagined him in a coffin.

'I'll go now,' he said. 'I might be going to join the army. You know, peace-keeping in the Middle East and stuff.'

'Sure, get a medal,' June said.

'Time to take off,' said William.

Wayne walked to the door and nodded. 'They killed the river. And then that fella's death. It's all over. I'm stupid. I regret that night, and every moment of my useless existence.'

William gave him a little shove, and Johnson was out the gate and slammed it behind him. June watched through the window. He seemed devastated all over again. He threw the cake into the geraniums and the cream slid out of the box and ants crawled into it.

June walked out of the house and stood behind him, placing her hand on his shoulder. She stood there for a while in silence. 'Nothing you can say will ever make it all right,' she said eventually.

'It's about respect for people. Our people and what they feel about such deaths. It's not the first and won't be the last Aboriginal death in custody. I was there at the Royal Commission so many years ago, and I was very young, and I went with some Aunts up to that twenty-third floor to listen to every family testimony. And it was harrowing to bear witness to the pain, but they needed it, just like now. It isn't about one person. It is much more, who we are.'

Wayne nodded, but nothing came from his grim mouth. Then he said, 'I don't agree with what's happened to your people, you know? Us whitefellas stole land and killed Aborigines and drained the rivers and infested the dry landscape with rabbits and erosion, allowed cattle to eat every bloody thing, and called it economic progress, it's not right.'

'Time to go, dear. But first, what's happening with the case?' She stared directly into his eyes.

He blinked. 'Doubt has been cast on the investigation's findings and on the integrity of someone in senior police.' He was attempting to look as if he was experienced.

'You mean Sergeant Blackett?'

He nodded.

June touched his arm. 'We want to find out what might link the two deaths. Determine who can be trusted. I told Duckie that. She needs some closure.'

'That would be police business, not for an amateur.'

'You lot are the amateurs. Hopeless how you handled the first death, and worse now. I will get you the proof. You've got blood on your hands.'

June almost spat at him as he drove away.

***

Later in her room, June laid out her new listening device and practised recording. It was a wall microphone with an amplifier and digital recording device which she bought on the internet on a spy website. It was very exciting. It was amazing what she could hear through the wall. The 'spouse-buster' microphone amplified everything by a hundred times, and she could record it to play to police.

She was going to make the journey out bush and have another conversation with Bam Bam. No problem – she would just ask him if he was the murderer. She would wear her warm sweatshirt.

She drove back to Bam Bam's caravan after dark and carefully placed the device under the aluminium window ledge. It was a cool night but her face dripped sweat and her heart pounded as she squeezed some Blu Tack to stick the device in place. Then she placed some blackberry vines against the object to hide it, and walked to her vehicle to wait for him. It wasn't long before she heard the roar of his motorbike, and she felt a surge of fear and her throat dry up.

Bam Bam was drunk. He staggered from his bike into the yard. He must have seen her car; there was nowhere really to park, so it sat visible from the road.

He yelled out, 'I know you're fucking there, you black cunt! What do you want?!'

June walked with her phone as a torch and slid up to his caravan steps. He was sitting outside the door, smoking a reefer.

He stopped her by pushing his black leather boot towards her chest. 'No closer. You really are something! I have to give it to you, you've got guts. Now piss off before I forget my manners.'

June smiled and nodded. 'I'll go, Mr Jones, but I just wanted to say I know all about your dealings with the Ice Man and with Mr Wright.'

'You know nothing!' He looked completely at ease, even enjoying himself. She prayed the listening device was picking up the conversation.

She moved closer to the side of the van where it was hidden. 'You got away with murder, then? Why did you kill our Thommo?'

'Yeah, no one will prove anything. I'm sick of all this shit. He was nothing, and I belted him out on that site. I got paid, and that's one feral we don't have to bother with anymore. Are you next, sweetheart?' He tilted his head and blew a smoke ring towards her face.

She coughed. 'Why'd you do it?'

'He wanted to sell the stuff on for a bigger cut. You thought he was innocent? What's the word? Naive? No, mate, he was slicker than a ferret up an arsehole. He shitted up the water deals by interfering and threatening the corporation, and then, fuck me dead, he wanted to screw up my business deals with Ice Man. No worries. He was a marked man.'

'The police—'

'Will do nothing. Case closed. Those two local fuzzes are fucked anyway, with that other death. They're goners.'

'All right, sorry to bother you, I'll be on my way.' June sensed that she was running out of luck, that at any moment he might strike like a snake.

'Not so fast, darling, how about you suck me off first? But take your dentures out.'

'Thanks – tempting, but no thanks, I'm off.' She turned and ran off towards her car, sweat pouring down her back. She heard him laughing, and then heard the unmistakeable sound of a gun being cocked. As she started the engine, a blast of a shotgun rang out overhead. The smell of gun powder and the sickening shower of red sparks. So close. Her terror magnified with the thought of him finding her listening device. He might smash it, and it cost four hundred dollars.

Suddenly he was upon June just as she reached for the car door to lock it. A scream came from her throat, and he pulled her from the car, then yanked her hair and forced her down on the dirt. She felt him press his fingers into her neck like a vice, and she nearly fainted with pain. Light shot through her head; maybe he would cause her to have a stroke. He struggled behind her, trying to get his belt undone. He was going to rape her. Her elbow jerked upward right into his crotch. He yelped and released her neck. Then she was free, and she searched wildly for an escape.

She ran down the track into the dark bush. That athletic running training as a teenager paid off; she felt like Cathy Freeman. But her heart soon raced to bursting, and she could hear him lumbering behind her.

He hesitated at a fork in the path. She heard him swearing, and he yelled: 'Piss off, Aunty detective! You come near me again and I'll kill you like any other shithead.'

After half an hour hiding under a wattle tree, June crept back to her car. She drove off over the corrugations and shot out onto the dirt road. She ticked off in her mind the fact that the experience had been her fourth escape from a rapist or basher. At that moment she hated all bad men. If he followed her on his bike she was done for. She imagined him as the crumpled heap on the bush path, a victim of his own gang. Then she pushed the image away; she was never ever going to be as cruel and violent as them.

She would have to wait a few days to collect the listening device. She prayed he would not find it first. She had him recorded; surely now it was over. Except she would probably have to play it back and also listen to him wrapped in a blanket masturbating over porn sites, farting and burping as he cracked a dozen beers. Boys will be boys. It was all in the name of justice.

* * *

Late that night, Johnson flicked his curtain closed and stared at the spattered ceiling above his single bed. He clenched the bedspread. If he didn't get up soon he might never get up. He traced with his eyes every crack in the clumsy paint job overhead. It was a map of a foreign country. That transfer was coming soon. Charges dropped. Or he might move to Thailand and find a nice respectable girl. Someone to love – who might give sweet kindness to a husband. His life was a daddy-long-legged spider moving across the ceiling, not knowing which way to run. His father had told him he was useless, but now it was time to rise from his bed and find her. The rain beat on the glass.

He dressed in black and put on his Kmart hoodie. He meant to walk the cold night away. She had blocked his phone number. Even in rain and thunder, Johnson drove through town. It was a shameful old vehicle, all that he could afford. He imagined the girl somewhere, a hunted prey. He willed himself to lift the silken veil. Really see new love. The blinding darkness split in half by lightning, bright as a wild white horse.

CHAPTER 37

25th September

June handed her niece a cold lamb sandwich with pickle. Arana did not eat it. The unseasonal warm rain continued for days, and Arana leant against the window, breathing on the glass and tracing Steve's name in mist.

Arana's bed looked cold and repellent, the sheet bundled at the end. She took her ball dress in her palms and inhaled Steve's smell from the lace bodice. There was a long hair from his head and it curled around her fingers. Her chest cut open and lay next to his pale body. She remembered his arching, lithe body against hers. Remembered the light of his political fire, his intelligence, his smiling, loving looks. Beautiful when he studied her. Not plain, awkward or shy, but radiant.

The young girl turned her head away as her mother checked on her. The door opened with whispering outside. Arana might run into the cool night and fly away. She glanced at her underwear, the white bra, and imagined him trying to unhook it, laughing out there on the desert floor. She could never wear it again. It had been in his hands. Hide it, so the overzealous aunty couldn't put it in the washing machine. Through the open window she heard sirens and distant shouting. Steve was drifting in the mist above her as she moaned on the edge of sleep and wakefulness.

Arana walked down the hallway and traced fingers along the wooden boards. Looking at her phone, she hesitated. He used to answer a call at his boarding house in Sydney with a shout across the hall. *A gorgeous dubai was looking for him.* She smiled at the phone as if he could still hear her.

On the coffee table next to Jack's empty beer can: a letter addressed to her in swirling handwriting. A fright went through her. She opened the envelope: a pink heart-shaped card. It was from that gunjie Johnson. He had drawn her a love picture with roses and blue birds. His signature was a flourish with a heart pierced by a spear: Wayne. She needed to vomit. Merle stood half-awake beside her.

Another card. This one was black and had a red rose on it. Sorry for your loss. Signed: Cedric Jones. Whoever he was?

'What's wrong?' Merle said.

'What can I do? Can't sleep. Can't bear it. He was sitting on that chair a month ago. Steve is everywhere I look. He's gone and that stupid copper has sent this card.'

Dogs howled.

'He's a pathetic idiot.'

'I'm dying,' Arana said. 'Don't want to be here. You repeat that it's okay. It will pass. But I don't want it to pass. To just get over it. Like Steve was nothing, like Thommo was nothing.'

'I didn't mean … You loved him, sweetheart. Suicide is no answer for us. We have to fight back, like our ancestors.'

An exhausted quiet came over Arana. Each new day to lighten the burden. The stupid love texts and dumb heart emojis polluted her mobile. The girl wanted to throw it at the copper's face. She breathed out. Steve would have laughed, her feeling sorry for him. Thinking of Steve as a past thing. Violent misery in waves, and her mind grew used to the unthinkable; acclimatised. A changing river, one moment dry as dust, next flooding. The fresh would come. She ate a crisp white roll with ham. Steve had liked ham with tomato.

She had been seeing Steve for months before the Knockout, but kept the secret. Her golden lover. Time stalled. There had been moments before he met her parents when she should have told them, but they were overprotective, and she enjoyed a secret life. Meeting him at the university with no local Koori to catch them and blab to her family. That Koori Murri grapevine was lightning. She imagined the tree-lined courtyard where they had lain in each other's arms on the cool stone, and the university library where she studied for HSC. They had gone to the Manning Bar, even though she was too young. So sophisticated: a glass of champagne from a long-stemmed glass.

In Wilga, the rainfall was earlier than normal. The news of upstream water rising filled the town with expectation of a bursting river. Everyone knew the bounty wouldn't last with the cotton farmers pumping it into their reservoirs.

'I'm going for a walk,' said Arana.

'It's raining and dark,' said June. 'Don't let life slip by. Every day has a meaning. Make it count, darling girl.'

'The rain's stopped. I've got to go out, this house suffocates me. Let me go!'

June nodded, and Arana slid past her into the night.

* * *

Jack had been drinking with mates and smelt the storm from under the bridge, where he was having a quick nap. He sat up and banged his head. He knew he might drown under there. A cold flash flood, a raging Noah's ark, and the end of the world. He smiled because the riverbed would suck rain in, and maybe the animals could survive and the fish return. Nature was a real beauty at that stuff.

'Baarri-y rise up, you river!' he shouted into the sky. 'Gali Balalaa come back, bring lightning. Muuri come down fast. Fill up the dry land!'

His people were rejoicing; they would hold a ceremony. The Ngiyampaa survivors. Yes, maranguka giving hope to people. He could rise, a true spirited man. No more stopping their river. The serpent raged, washing away the irrigators.

Staggering along the road, Jack felt thunder and flashes deep in his bones. It shook the ground, and he was certain any moment it would strike for him, turning him to ashes. He chuckled and didn't care. 'You call this electricity?!' he shouted to the sky. The electroshock treatment he'd had long ago had been much worse because it had been right deep inside his skull. A skeleton jerking. Freedom: his face in clouds, and he might levitate, free from the memories. Jack reborn, a new man, no longer mad.

* * *

Arana walked along the sodden country-town street towards the rising river and levee bank. She heard rushing water charging into the gully and imagined the joy for the Barkindji and Murrawarri people downriver. The earth turned to sticky grey mud as she stretched her arms out to the rolling grey clouds lit up by a drowning moon. She passed the lights of flickering television sets in people's houses, glimpsing them sitting on their night and day lounges with their mouths open. So many dead fish. But when someone ran into her, they turned away, as if the death was contagious.

By the stream the mist rose and the warm rain plonked on the surface. It was flooding and Madukka the great serpent was awakening. Like the corroboree danced above the fish traps, a long line of dancing ochred women holding the long human hair rope to dance life for the serpent river, to sing it back to life. The old Elder in red striped paint clattering the rhythmic boomerangs and singing songs forty thousand years old.

The girl spread out on the muddy bank, her arms outstretched with rain pelting her eyes. Arana heard the roaring river coming alive. She faced the black cloud and shuddered. Here it comes, she thought. A howl flew up the curving torrent. She was a victim tied to railway tracks, unable to lift her hands from the ropes tightening her wrists. It was the moment to die: simple, waiting, the serpent would engulf her. The trees blackened and sang, leaning into a wind that tore the dead branches apart and flung them into the dark. No man on a white horse raising her up and galloping away. No bloody prince. She gazed at the storm face. She might drown by just opening her mouth. The roar grew louder, and she wished to block her ears from the terrifying sound. It was a levitation, uplifting on the howl into the arms of Steve. A coolabah tree broke in half and goonge poured out, released. Her eyelids fell; expecting to die. Then, as she peeped out, a shimmer of purplish light shone on the track. It was time to stand up and walk towards the riverbank.

* * *

Up on the riverbank, hidden in dark cool trees, Johnson stood as water dribbled onto his body. He sobbed and thought they had visited all the plagues of Egypt upon him. He needed to float away too. Wait for me ... His shoes sank in a puddle of cold black mud. He held up his arm and waved. So lame. Nothing, of course; he was stupid, again. He picked a pimple on his neck and sighed. He wanted redemption. Beg her, then she would forgive him. Then become a greenie, yes. Her river would return. Johnson walked back along wet streets to his room to get into bed and shake with a sheet pulled over his wet head like a shroud.

* * *

Out of the darkness a figure emerged from mist. Arana recognised Sergeant Blackett checking the riverbank levee with a police torch. He kicked at a pile of sandbags and sat on one to sip from a whisky flask. It was bound to happen, this unfortunate meeting. This date with facing questions of life and death. She stood still, breathing fast by the casuarina trees. She noticed his easy way of wearing the cow cocky clothes, passing for a farmer in a blue lapelled shirt. A moment of panic. Arana might hide behind a tree or run like the devil.

The rain trickled on her face, and, feeling powerful, she willed herself to stand still until he found her. *Come to me, come to my fiery self, make me an avenging angel, come to my rage.* But impulse propelled her to squat in the sand. She made herself small as an echidna. Arana pressed her knees forward and rested her head in prayer. Every movement pulled him to her. Every step closer would look to spell out what she needed to discuss. Their history spilt on the earth. The resurrection of terrible deeds.

He sang to himself. What was the song? 'The Dock of the Bay', for God's sake. She wished that he might tumble forward into the grey slush. It would consume him.

Then, with a miraculous vision, the sky cleared, and Arana looked up. Blackett saw her, looked at Arana in her wet clothes, and he stood up, cold and shaking. Their eyes met.

'Hey, you, miss, you right? Be careful! God, the river's high … I've been lifting bloody sandbags all week.' Blackett stumbled, with the torchlight shining on Arana's face.

The thunder cracked the sky, and lightning zipped across it. The serpent alive and angry.

'Piss off,' said Arana.

'What's a young girl doing here? It's not safe.'

She imagined him dissolving in the rain, not real. Her eyes were blurry with flood, and she shook her curly head and the droplets showered them. 'I need to ask things.'

He stared with disbelief, and looked confused, but she stood tall and calm, then laughed at the absurd meeting. As though she could clap her hands and he would disappear. A magician's trick. Yes, he was actually just flesh and skin and fat.

'Come over here, I won't hurt you,' said Blackett.

Hurt you? she thought. No, not hurt you. Just murder in cold greasy gore. She imagined his hands soaked in it, with a sock in his grip and the stupid copper Johnson simpering beside him, an idiot.

She stood slowly and walked away into the darkness; her back swayed like a siren's. But she sensed him close behind her, following and wheezing. The hair on her neck prickled.

'Where are you? Hard enough to see a white girl, but you Black ones ... Hey, smile so I can find you.'

She turned and walked back towards him into the torchlight, and she stared at his balding head, his bulbous pink nose and his sad flabby body.

'Don't walk away ... It's dangerous,' said Blackett.

She imagined him with his pants curling around his ankles, grunting, fat stomach, fingers clutching at dark buttocks.

'I recognise you from somewhere,' said Blackett as he loosened his belt. 'You're not from round here, are you?'

Now her self-assurance was fading, and it was being replaced by rising fear. 'You don't remember me, do you?' she hissed.

'Hey, I can't remember every girl.' He wrapped his arms around himself to disarm her. He was just a gentle old man working on a levee bank. Then, in an instant, he knew her. An alarming thought. A shot of panic like heroin.

'What happened that night?' said Arana.

'Oh yeah, June's niece. I got post-traumatic stress disorder. A nervous breakdown. I don't have to justify my actions. I protected that constable. Showed leadership. Could've told the truth, but ... Lot of rain, I'll look after you.'

'You looked after Steve,' she said, and she slipped in the grey mud and it stuck in clumps to her shoes.

'What? We did our best. It's hard, the police. Hey, don't be like that. Be careful! It's wet.' He laughed.

She watched him reach out with a clawing, beckoning gesture as he slipped also. A huge bear in a Disney cartoon, or Foghorn Leghorn.

'Shit!' he said as he wiped mud from his shoe.

'Why don't you tell me the truth? You lied on the report,' she said.

But he was closer now. He tramped in the mud, and he was upon her; she could smell his sweat and Lynx deodorant. Arana turned away in disgust. A pathetic, broken man. She leant backwards and avoided his grasp as he fell again in the slush, and she watched his hand lift towards her arm to steady himself.

'Don't do that,' she said, and looked at his hand and stopped him with hands raised in supplication. Afraid as her strength dribbled away. She wanted to run.

His hand stroked her face.

She shuddered.

'I won't hurt you,' he breathed, and his gaze was gentle and serene, as though he would blow a kiss. He wiped a tear from his eye.

'I want to ask you something. Tell me what happened to Steve,' she whispered.

'Who? Sorry, darling, aren't you forgetting I'm a copper? I can't divulge any investigation. Look, pet, go home. You're soaked.'

Her breathing became heavy, and he grabbed her hand as if to pull her closer.

'You lied.'

'Look, miss, I wasn't responsible for that death, and I have suffered enough as a consequence. It stuffed my health, can't bloody work.'

'You should own up, be a man,'

'It was your mate, William's friend, Johnson. He did it.'

Her mouth gaped open, then she shook her head. 'No, you're lying.'

He had her by the arm, and she wrenched away with a violent pull. Blackett let go of Arana as he stumbled backwards on muddy stones and fell. The sky rang with lightning, as purple and white flashes lit up the river and the trees bent in the wind. There was a howl from the water and wind like a devil-devil waking. Steve's ghost was a misty shadow, and Blackett looked up. It hovered above him. A ghostly fear. The sergeant looked at the white river rising and knew it might overflow the riverbank.

'Shit, we better get back up!' said Blackett.

He took hold of a stack of sandbags, heaped them on top of the levee and rushed to stop the pale water breaking the bank. He seemed to have forgotten Arana; his attention was on the stream. A seething cold flood would sweep into Wilga and there was no one to stop it. Water roared downstream.

Brown water tore open cracks in the bank. Blackett had a sandbag in front of his chest, and the weight tipped him forward. He worked in deep oozing mud. A steep levee bank gave way beneath him and he struggled to pull himself up the wall with fingers clawing the black slime. It stank.

Arana froze.

Blackett slipped further backwards into the muddy quicksand as he grasped at tussocks, but they tore out in his fists as he slid, grabbing at anything. The water rushed white behind him and he disappeared under the raging torrent. A gurgling noise. His balding head rose and sank again with panicked eyes.

Arana stared at the grey water. The sound of the river became deafening. 'Gran said not to swim across the middle: something lives there. It eats dogs and men.' She walked away.

Blackett rose in grey water, choking. He attempted to swim in the tearing torrent, shouting, 'Help, help me, for Christ's sake! Get a branch and pull me out!'

Arana looked coldly at him and hissed.

The water bubbled white and cold over his pink scalp. He sank.

She hesitated, then watched with grim satisfaction as his pale hand wavered in the flood, and she walked away. She shivered; her jumper soaked in the deluge.

The serpent took Blackett away as it tore through Country. That gushing, swirling snake, the Madukka, Waaway, Ngatji from the Milky Way, drowned him in its rage and tumult. It mapped the path of the Old Man Crow who chased the two sisters, and in their footsteps magic places made the land forever Aboriginal land. Those spirits possessed them all. Bida–Ngulu lived in the sky and spoke out like thunder.

***

June, Merle and William searched for Arana for a long time. They drove around and around the town and in teeming rain, feeling terrified that it might sweep her away. Then June saw her by the levee bank, and Arana strolled towards them, looking quite happy but dripping and frozen.

'Arana, are you okay?!' June yelled as Arana walked towards them. 'We were that worried about you. Get out of the rain, you'll catch your death.'

William rushed to take her in his arms. He was nearly weeping. 'We love you, our daughter. This flood will save us all,' said William.

'I'm good, I just wanted to think and be alone by our river. It's flooding, and that's a good sign. Don't worry, I will be fine.' She smiled, and they laughed in relief.

'People who kill themselves have given up the fight,' said June.

'Don't worry about me. I have to study for my high school certificate, so I better go home to Sydney. I will go north one day to meet Steve's family. I owe them that. He is here around us now; we can sense him, right?'

Merle helped her exhausted daughter into the car, where she curled up on the back seat and fell into a deep sleep. June looked over at William and he pushed a sandbag into the broken levee.

Just then two council trucks arrived, full of men and women in yellow fluoro vests. June cheered them on. A State Emergency Service vehicle pulled up. The men threw bags from the trucks while the grey flood rushed to peer at the levee bank.

A council foreman yelled through the deluge, 'All you CDEP workers and council employees, we're out here to work, it's not a council bludge job, we need to shore up the levee bags. It's life or death. Put your backs into it.'

The workers filled sandbags, and they shouted orders through the rain and hurried to build a barrier to halt the flood. William ran over and grabbed a sack. They nodded at him, and he began filling it. June rolled up her sleeves and helped carry bags. She wiped the rain from her face, and Merle took hold of another sandbag, and between them they dragged it to the dam bank. They looked like they were enjoying it.

June watched as a worker ripped off his shirt to hold up a bag as Fay Pascolini added a scoop of sand and patted it down. He heaved the bag onto his broad back to carry to the levee.

'That you, Arthur?' shouted June. 'I didn't recognise you with your clothes off.'

'Had your chance!' He waved and mopped his face with a handkerchief.

'You've got nice muscles,' she said.

'Sounding sexist, June,' he laughed, and shovelled another sandbag full.

The rain drenched their bodies.

'Keep shovelling, Arthur,' said June.

'Black and white together, we'll save this town,' he said.

Merle laughed and dragged bags across the mud while William stacked them on the levee wall. After three hours they stopped the torrent, and the group stood by the wall, patting it down.

'Okay, lads and lasses, it's a good job. Back to the hall for a cuppa,' said the foreman.

* * *

Days later, the family heard on the wireless that Sergeant Blackett had drowned in an accident.

June called out across the lounge room, 'Hey, someone on the radio says he's a brave hero. The town councillor, Mr Wright, has announced that he wants to erect a plaque to commemorate the police officer's heroism, or perhaps a nice stone monument. There's a Sergeant Blackett memorial Facebook page to raise the money.'

The house was quiet. No one shouted in jubilation or showed any emotion. June lay on the lounge and wrote out a notice to print.

She called a community meeting about the murders, and invited the new police officers who had arrived in town to replace the old ones. She put up signs in public and on the internet; she was ready to explode the case open. She was going to tell the whole truth and nothing but the truth.

On the night of the meeting the Murri community filled the other seats. All eyes were on June. William nudged her, and they watched Aunty Mary hobble in on her walking stick and wave to everyone. Like a star. Like a trusted Elder.

June put her head down. She looked sideways, and noticed that outside the building two police officers stood near with their weapons. They terrified her.

She stood up on the stage, adjusted the microphone and began. Her voice shook. 'I'm a trained investigator, and I'm going to tell you who is responsible for what happened to two of our mob.'

She was tall and proud. Now it was time for justice.

'First, we have evidence to give to the police, and we hope they will make the correct arrests.' She coughed, and the paper in her hands shook. 'We know who killed Steve: it was the two Wilga police officers, and eventually we will hope to get justice. However, the young man Thommo was murdered in cold blood – he was beaten by a warraaykuuya, a bad white man. We have given police a recording of certain people speaking about how the Ice Man ordered the murder. I can't divulge the name of the hitman, but he is here tonight.'

Yelling filled the air, and Murries banged their chairs and stamped their feet. She held up her hand until the room was quiet. Merle stepped up behind her and pressed against her side to steady June, and she continued.

'He says we can't prove it, but we have a recorded message on a phone where he admits to being involved. We have the murder weapon. We have the motive, and, no, it wasn't a drug deal gone wrong. Thommo loved his wife Duckie and their only son. He wanted to make a little cash to help with medical bills. He was an activist who cared for our river, and was hot-headed and brave enough to stop some pumps that were sucking the river dry. I salute his bravery! A man followed orders and killed our cuz out there on that lonely rock pile.'

She looked over at the Elders' group. Some sobbed.

She took out a tissue and rubbed her eyes. 'The man who did this fights hard, uses his shotgun. We all have targets on our back if we tolerate these dealers in our town. We can't keep living in fear! Say no to the code of silence.'

William got up and stood next to June. He crossed his arms and pushed out his chest. 'Tell them, sis.'

'We have evidence that Ice Man paid Bam Bam to do it. For control of the drug market. He had a motive: Thommo was foolish for getting into drug dealing,' said June.

Duckie had been hiding at the side of the hall, and everyone turned as she screamed. 'No, it's terrible, but! My poor Thommo! Oh hell, I'll kill him, but!' She leapt out of her seat and attempted to attack Bam Bam, but he put out his arm and grabbed her by the wrist. She stayed rigid with rage.

He laughed. 'Ice Man ordered the hit, threatened me. I had no choice.'

The police moved towards Bam Bam as he ran from the hall. The bikie turned his head and gave an 'up yours' gesture with his tattooed arm, and rushed into the street.

CHAPTER 38

1st December

Months later, June closed her office in town, took down her Certificate III and packed her clothes in a suitcase. She sat in front of the mirror, brushing her hair. The brown grass was turning to dust around the street, and some roses struggled behind fences. A flock of galahs were pecking at seeds along the edges of the road. The weather was hotting up. The sky was blue, but promising dark clouds were in the distance; it looked like rain again. People in the town had been dancing in it.

June walked her suitcase down the sweltering street to the bus stop. Santa pictures glittered in the shops' windows, fake Christmas trees covered in fake snow. A sweating Aboriginal Santa drank a cold orange juice outside the IGA supermarket. She wondered if Australians would ever get off the boat from Europe.

A few weeks before Christmas and there she was, sitting in the shade, counting her money. Getting away. Nearby, some Murri kids were doing wheelies with scooters on the grass near the empty emu cage. The air was warm and fresh, full of boronia perfume. She ate a KitKat, carefully folded the wrapper and put it in her bag. She checked the timetable on the notice board and saw a peeling 'Save the River' sign on the back of the post. So much had happened since she had glued it there with Merle. She thought about the lads, like Thommo, who joined every local environmental protest and chained themselves to an ancient coolabah tree. A sacred tree. Thommo, who admired non-violent direct action when he locked himself with a heavy chain to a water pump near a cotton farm. And fearless Steve, who was the future of Aboriginal law makers, floating above her in the blue sky, unable to rest. Such a bitter waste.

She took a Travel Link bus to Sydney. The rest of the family would follow by car in a few days. She sat in the nice air-conditioned bus and drank a bottle of peach juice, and they stopped at the garage for refuelling. People got out to buy hot chips and chocolate milk.

June walked around in Redfern, and saw the faded mural celebrating the Redfern Blacks footy team in the 1970s. They looked so proud. It was being demolished to make way for a multimillion-dollar Aboriginal Housing Company development. She sighed.

She leant against the wall, sending her mind to that cool green river, luscious and fast-flowing, bringing clean flood water to the heart of Ngiyampaa and Barkinji country. The Ngurrampaa country called out. Waaway made the river and he swam and surfaced through clear green water, his eye sparkling like a star. The land would drink in the power, and once again the river was the serpent's path, the power of Madukka, Ntji, Waaway and Karupuli. The Wangkamara, Murawari, Gamilaraay, Yuwaalaraay, Wiradjuri, Jitajita, Kureinui, Maraura, Bakindji, Parundi, Ngiyampaa, and Ualarai: all had drunk water from the Darling.

Acknowledgements

ACKNOWLEDGEMENT OF COUNTRY

I pay my respects and acknowledge the presences, places and peoples of the North-Western New South Wales Indigenous lands and communities, particularly those of the Gamilaraay and Ngiyampaa nations. I acknowledge Elders, past, present and emerging, and recognise that they always have been and always will be caring for Country.

NOTE ON LANGUAGES AND PROTOCOLS

The Indigenous languages quoted in the novel are the languages of the Darling River and lands beyond. Gamilaraay and Ngiyampaa words are used in the context of Aboriginal Murri community speech. These words have been gathered through workshops where the custodians have contributed their knowledge. I thank the contributors of language and honour their custodianship of such material. I pay my respects to the knowledge holders of these languages and the community people represented, past and present.

THANK YOU

I thank UWA Publishing and Eleanor Hurt and Kate Pickard who worked with great passion to bring this novel to publication. I also thank the wonderful proof reader Emma Schwarcz.

Madukka the River Serpent is a novel set on the Barka (the Darling River) in northwest NSW. This novel is an Aboriginal female detective story that deals with the drying up of the river around 2020 onwards and the human beings who are responsible for the river's destruction. Water theft and political intrigue are central to the fictional story, and the Aboriginal spiritual and ecological life of the river are at the novel's heart.

ACKNOWLEDGEMENTS

I thank my agent Sarah McKenzie of Sarah McKenzie Literary Management and the excellent editor, Kate Galsworthy. The Australia Council provided much-needed support. My thanks to FNAWN (First Nations Australian Writers Network) and to readers: Geraldine Starr and Virginia Fay.

This novel developed from my play 'Gunjies' (police) written in response to the 'Royal Commission into Aboriginal Deaths in Custody' in Sydney from 1987–1991. This play was produced at Belvoir St Theatre in 1993, the International Year of Indigenous People. It received an AWGIE nomination and Highly Commended Human Rights Award. 'Gunjies and Black Mary – Two Plays' were published by Aboriginal Studies Press in 1996.

'Gunjies' was written while I was a Lecturer in Aboriginal Communication and Drama at the Aboriginal Education Assistants Program, University of Sydney, and at the Aboriginal Education Unit, Liverpool College of TAFE and Tranby Aboriginal Co-Operative College. The play received support from the Australia Council for production at Belvoir St Theatre and was developed with assistance from Playworks Women's Writing Workshop. 'Gunjies' has also been produced at the West Australian Academy of Performing Arts in Perth, West Australia.

In the writing and research process for the play, I was given support and Indigenous knowledge from a large number of Aboriginal people: during drama workshops at the University of Sydney we created the characters of June, Merle, Arana, Uncle Jack and Barney (William). I give thanks to many of those Aboriginal students and teachers for their help over thirty years.

Australian Government | Australia Council for the Arts